SOTHIS PRESS

Sothis Press aims to completely reimagine the public domain book with an emphasis on universal access to captivatingly designed print editions. We believe that technology will enable every single book ever published to be perpetually accessible in beautiful print form. Sothis is beauty in ubiquity.

www.sothis.press
1 Waterhouse Square, London, ECIN 2ST

Hand-Book for Lumbermen

Saw, Tool, Steel File Works, Philadelphia, Pa

Henry Disston

SOTHIS PRESS

ISBN 978-1-0365-0392-5

Hand-Book for Lumbermen
is in the public domain.

This edition is published by
Sothis Press © 2023. The company
is registered in England and Wales.
The Company Number is 14900389.

HENRY DISSTON & SONS'

[INCORPORATED.]

HAND-BOOK FOR LUMBERMEN.

BRANCH HOUSE,
Chicago, Ill.

TRADE MARK

BRANCH HOUSE,
Louisville, Ky.

KEYSTONE

Saw, Tool, Steel & File Works,

PHILADELPHIA, PA.

PART I.

CONTENTS.

PART I.

CONTENTS.

INTRODUCTION

In 1840, when Henry Disston commenced the manufacture of saws, the lumber business, which has since assumed such vast proportions, was in its infancy. It had not been brought to the science it has since attained; saws were not required to stand the tremendous speed and feed which is now demanded of them; lumber was plenty, and the thickness of the saw was not a serious question; since then mills have been introduced that drive the saw at a tremendous speed and the feed has been increased proportionately, consequently, saws must be made to stand this increased demand upon them. Still, there is a limit to the strength of steel and saw, and while we do not claim that every saw we make is perfect, (as it is impossible to discover slight imperfections in the body of the steel plate, which in spite of all care will at times exist), we do claim that if saws of our make are treated as we suggest in this pamphlet, they will meet the requirements in almost every case.

Owing to the great difficulty experienced by us in getting steel suitable for our wants, in 1855 we commenced the manufacture of saw steel, securing the services of men who had had long experience in manufacturing it.

The reputation which our saws have gained and the frequent orders received, which specify that Disston's steel must be used, is a guarantee that our steel is of superior quality, and we claim a great advantage in being able, from the fact that steel and saw works are connected, to readily detect any faults which may arise. One can see how important it is for us to use the greatest care in the steel works so that the saws will not be condemned after we have expended so much time and labor upon their making; therefore, the quality of steel used, the perfect system of grinding, tempering, hammering and testing which we follow, with the great experience we have had in the business, has gained for our saws the reputation which they are justly entitled to. But notwithstanding all this care and attention on our part to produce a perfect saw, we are compelled at times to suffer from a want of experience on the part of the men engaged in the mill business, who persist in running a saw beyond the limit of endurance, and the improper methods used by some to keep their saws in order. We are always ready and willing to replace a saw faulty in either steel or manufacture, but we do not think that we should be expected to replace a saw which has been subjected to unfair usage.

Mill-owners and sawyers will find it greatly to their advantage to keep this hand-book in a convenient place for reference, for it contains a great deal of useful and valuable information for them.

SUPERIOR TO ALL OTHERS.

Having had over forty-seven years' experience in manufacturing saws, we feel justified in saying that our goods are superior to all others, from the fact that we have, by constantly experimenting at great cost, arrived at a state of perfection in machinery for manufacturing saws which can only be attained by years of constant application and watchfulness. Disstons' Saws are used wherever lumber is made, and are the criterion of excellence by which the merits of all others are judged.

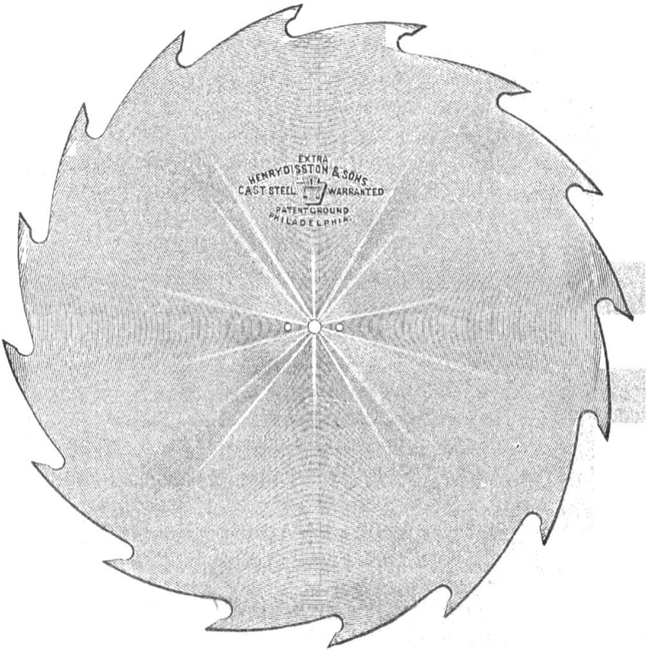

Parties ordering saws from us will save themselves much trouble by exercising care in making out their orders; we furnish blanks to facilitate this, which can be had on application.

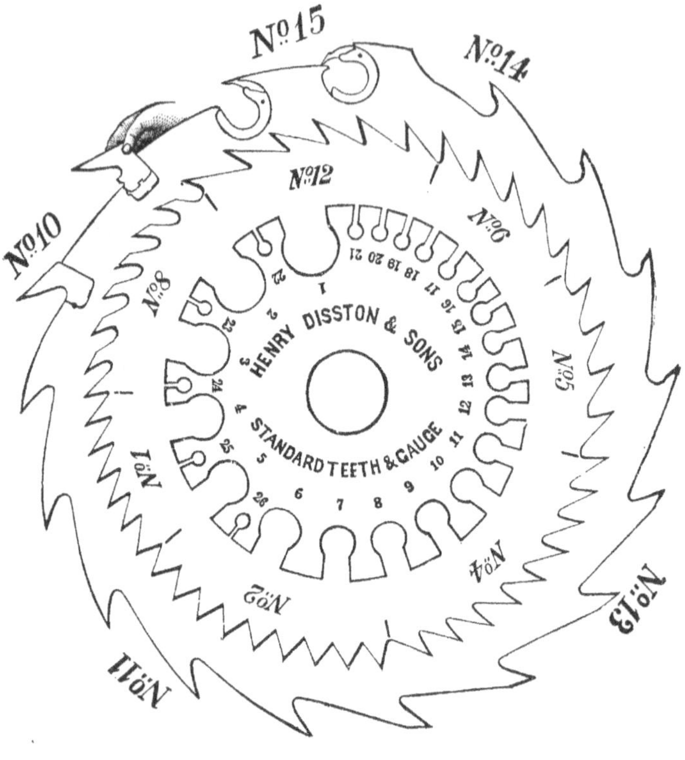

The above illustration represents our various styles and Nos. of saw teeth; also our standard gauge. By consulting it a person will be enabled to inform us the size and style of tooth, and also the gauge of any saw he may desire.

STANDARD GAUGE.

Gauge No. 4,	$\frac{1}{4}$ inch	scant.		Gauge No. 9,	$\frac{5}{32}$ inch	scant.
" " 5,	$\frac{7}{32}$ "			" " 10,	$\frac{1}{8}$ "	full.
" " 6,	$\frac{3}{16}$ "	full.		" " 11,	$\frac{1}{8}$ "	scant.
" " 7,	$\frac{3}{16}$ "	scant.		" " 12,	$\frac{3}{32}$ "	full.
" " 8,	$\frac{5}{32}$ "					

BLANK FOR ORDERING MULAY, MILL AND GANG SAWS

From Henry Disston & Sons, Philadelphia.

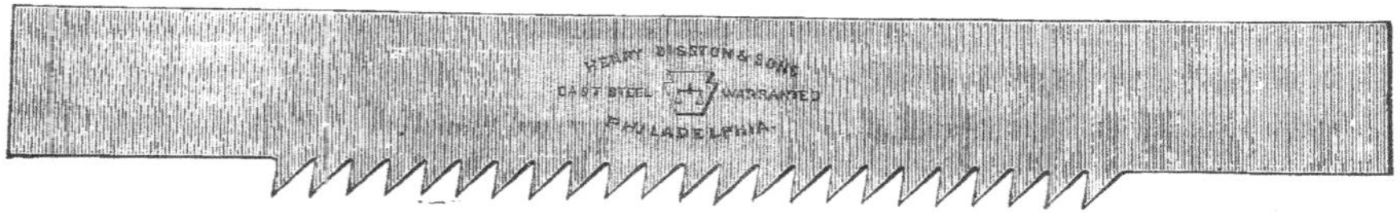

Length, ...

Width, ...

Gauge, ...

Space of teeth, ...

No. of teeth, ...

Hook of teeth, ...

Distance from bottom of Saw to point of front tooth, ...

....

Width at ends, ...

If tabbed, give exact distance between tabs, also thickness of Saw and tabs when finished, ...

Kind of tabs wanted, ...

ALL SAWS WITH OUR BRAND FULLY WARRANTED.

Branch Houses:

Randolph and Market Streets, CHICAGO, ILL.

850 and 852 West Main Street, LOUISVILLE, KY.

HENRY DISSTON & SONS,
PHILADELPHIA, PA.

DIRECTIONS FOR SHIPPING.

By .... *Express.*

By .... *Steamer.*

By .... *Regular Freight.*

Signed, ...

RIGHT-HAND SAW.

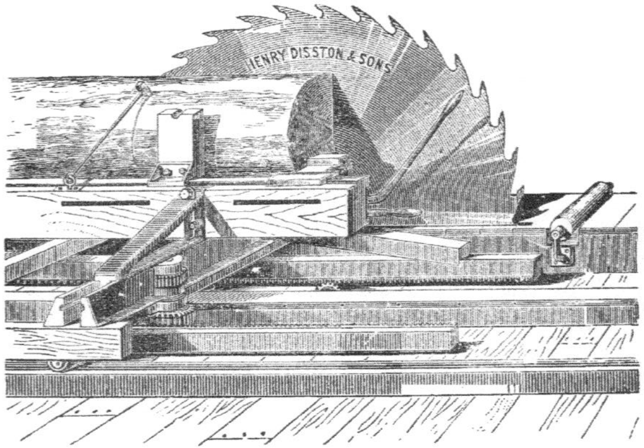

LEFT-HAND SAW.

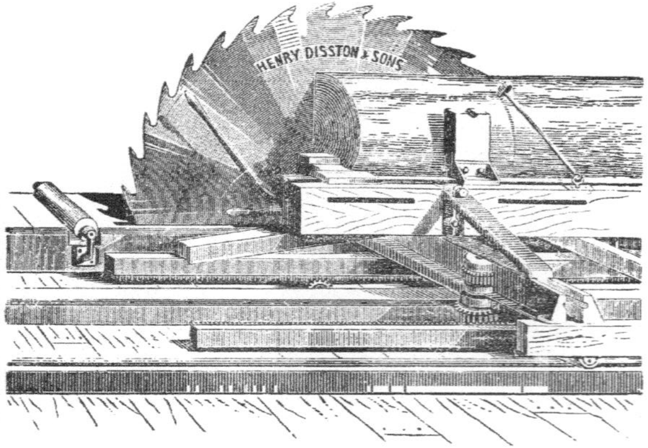

When ordering Circular Saws, the following directions should be explicitly given:

Diameter in inches; thickness or gauge at rim; thickness or gauge at centre; right- or left-hand (see engraving above); number of teeth; kind and No. of tooth, as shown on engraving (see page 9); size of mandrel-hole; size of pin-holes; distance between pin-holes from centre to centre; greatest feed at each revolution, in inches; kind of lumber to be sawed; number of revolutions per minute. When ordering bolting saws, state whether rip or cross-cut.

N. B.—All our stock saws forty inches and larger in diameter have 2-inch mandrel-hole and 5/8 tug pin-holes, three inches from centre to centre. If wanted different, please send full pattern of holes.

Hints to Sawyers and Saw-Mill Men.

A GOOD MILL.

In the first place you want to make sure of a good mill; a few hundred dollars more at the commencement is a small amount when compared with the time and money lost by a poor one. A few weak points about it will sometimes consume half of the power, *but no matter how perfect your mill may be*, unless you have your collars true and your saw well-fitted and nicely sharpened, your neighbor, whose mill may not be as good as yours, will do better work if he pays strict attention to keeping his saw in order.

A GOOD SAW.

It is quite as essential that you have a good saw as it is that your mill should be good, true and solid. Our saws stand at the head of the market on their merits. In workmanship, toughness, elasticity and quality of steel, we claim Disston saws unequaled.

If you will write us when you are in need of saws, giving us a description of the work to be done, we will cheerfully give you the benefit of our experience in selecting a saw, the diameter, gauge, style and number of teeth which will be best calculated to fill your bill, and which we will guarantee to do your work satisfactorily.

Below you will find some of the causes which give rise to complaint against saw and saw-maker:

A saw that is *too thin;* a saw that *has not sufficient teeth for the amount of feed;* a saw *not properly hung;* one *not evenly set;* that is *not filed square or true;* a tooth that has *not enough pitch*, or that has *too much pitch;* a tooth whose *back is higher than the point;* that has *not a free chamber for the dust;* a saw *not properly balanced on the mandrel.* To balance properly, a saw must be perfectly round, must have teeth of equal size and shape and round gullets of equal depth, or one portion of the saw will be heavier than another, causing it to leap and tremble. A saw too thick in the centre will sometimes heat. *A carriage that springs; a carriage not properly lined with the saw; a mandrel that springs; a mandrel that heats; collars not true* also give rise to complaint. A gauge too long is frequently the cause of great trouble with bench saws. A very general cause of trouble is a *dull saw*, not only dull on the extreme point, but the cutting portion of the tooth under the points, as illustrated in Fig. 1, on opposite page.

The points of the saw-teeth are the only portion of the saw which should come in contact with the lumber. They must be kept sharp by frequent use of the file, and set by springing, swaging, or spreading when necessary, sufficiently to clear the blade of the saw nicely to prevent friction. As the points of the teeth do all the work, they speedily become dull and round, the sides of the points wearing away as well as the points themselves; take great care that the points are kept right; they can be kept so by using a Jumper or Upset; see pages 39 and 40. A good sawyer can tell when sharpening and setting his saw whether the temper is right or not, and by the use of a straight-edge can see if it is true and ground to his liking; a new saw should always be tried before and after the collars are tightened, to see that it hangs true; should it not do so, let us know at once where the fault is and we will correct it if in the saw, but do not in any case attempt to run a saw of which you are doubtful, for by so doing you might spoil the best one ever made.

SAWS OUT OF ROUND.

To remedy this defect, a piece of grindstone should be held against the points of teeth while the saw revolves and thus reduce or wear down the most prominent teeth; or, a piece of red chalk may be held against the points which will mark the teeth in proportion as they are long or short, the long teeth being reduced by filing; for when a saw has long and short teeth, it naturally follows that the long teeth will have the most work to do, thus bringing an unequal strain upon that part of the saw, which will cause it to leave its line, heat, and give bad results generally. The cutting of a circular saw should be continuous, and to be so must be *perfectly round*, otherwise you cannot obtain the best results. On the same principle, the tooth edge of band, gang, mill and mulay saws must be *perfectly straight.*

Neglect of any of the suggestions above given are sure to cause you trouble; we have had saws returned us as shown in Fig. 1 at D, not because they were bad when they left our factory, but because they had not been properly kept in order.

Fig. 1.

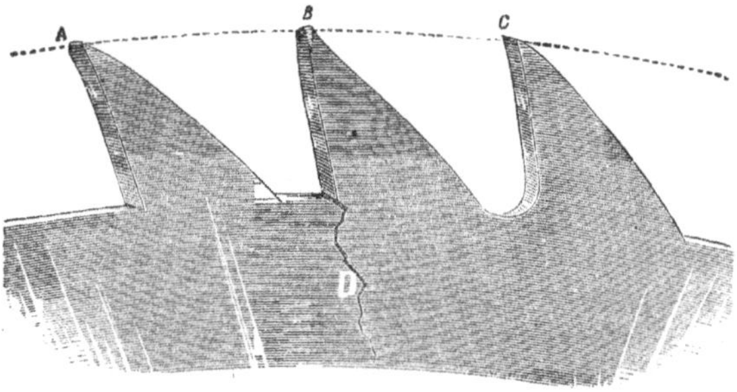

DULL TEETH AND SQUARE GULLETS.

Do not file square corners in the gullets of your saw, as it is very liable to break as shown in cut above, at D, particularly when dull or in frosty weather. Our *warranty is cancelled if these square corners are filed in your saw.*

You will observe in the cut on preceding page that in addition to having sharp corners in the gullets, teeth A and B are very dull; tooth C shows how the points and gullets should be dressed. The gullets should be kept rounded out, either with a burr gummer or a file.

Your saw-tooth becomes dull on the side or under the point in proportion to the amount or extent of feed; for instance, if your tooth takes one-eighth of an inch hold at each revolution, it will become dull for one eighth of an inch below the point, or more or less as you increase or diminish your feed. A few minutes filing two or three times a day will save ten-fold the amount of time and labor in running a dull saw, also making a saving in the power consumed, which, in some mills, is a very important consideration, and a heavy percentage in the quantity as well as the quality of lumber cut in a day, month or year. It is a mistaken idea that there is a saving by not taking time to sharpen the saw; it is as essential that a saw be sharp as a razor or plane, in fact, any cutting instrument, and when proud, or full and sharp, it does not require one-half the set or power on the same feed.

IMPROVED METHOD OF CASTING STEEL INGOTS, OR COMPRESSED STEEL.

Patented, December 8th, 1885.

From the time that saws were first made to the present, saw manufacturers have had the greatest difficulty in obtaining steel plates for circular saws that were free from flaws caused by cavities formed in the ingots while they are cooling. These cavities injure the steel very much, rendering it entirely worthless for saw steel, for when containing them, it gives the saw plate the appearance of being made of two pieces of steel not properly welded together and causes the teeth of that portion of the saw to split, spall or crumble.

It was impossible to detect all of these flaws in the ingot or in the plate after rolling, consequently saws have been sent out by all manufacturers that have proved unsatisfactory, causing an expense to both the mill man and the maker; an expense to the mill man in loss of time, and a dead loss of the saw to the manufacturer. We have always taken the utmost care in the manufacture and inspection of our steel and saws and have always headed the list as saw manufacturers, but in spite of our best endeavors, we were, until quite recently, unable to entirely overcome this difficulty; now we are happy to say that after years of careful study and expensive experimenting, our efforts have been crowned with the *greatest success of the age*, as our *method* of casting steel ingots and *process* of compressing same, entirely does away with all splitting, spalling and crumbling of teeth. This process is entirely new and original and has long been sought after by prominent steel makers both in this country and Europe; several patents have been granted for compressing and improving saw steel, but all of them have fallen short of the mark until our patent No. 331,856 was brought out.

This method not only does away with the splitting and spalling of teeth, but makes a hard, tough, elastic steel of the highest quality, which, with our improved and patented process of manufacturing and tempering, enables us to furnish our patrons with saws that for toughness, standing-up quality, uniformity of temper and general superiority, have never before been equaled.

Among the first saws sent out made from this new steel were three 64-inch by 6 gauge straight, which went into one of the largest steam feed mills of the South, working on hard yellow pine, and we have their word for it that in a run of nine hours, they cut 45,000 feet of one inch boards,, using only two sharp saws to do it with. Many other testimonials of the same nature prove what we say above, that our efforts in improving saws and steel have been crowned with the *greatest success of the age.*

HANGING THE SAW.

In this matter, too much care cannot be exercised. First, the mandrel should fit as tightly as possible in the boxes, so that it will run without heating. Secondly, we recommend, where flat collars are used, the steadying pins to be made with a shoulder (see cut below), for it frequently happens where any other kind of pin is used, a burr or bunch is raised at the corner where the pin enters the collar. This is avoided by using the pin as shown in illustration, the bottom of which is smaller than the top.

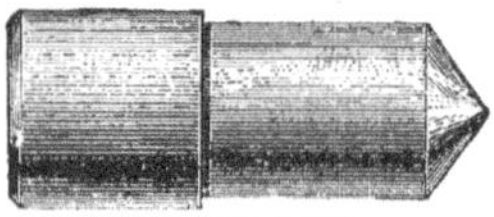

Steadying Pin.

SPEED OF SAWS.

This is a very important point for consideration, as a hundred revolutions, more or less, will always make a great difference in the running of the saw, yet, we know of many cases where the speed varies several hundred revolutions per minute; we can adjust the tension of the saws to overcome this variation in the speed if full instructions upon this subject are given us when ordering. But we advise a regular speed at all times; our experience has been that saws work better running one hundred revolutions below speed given in table than they will when working first one hundred revolutions or more above, than as many below proper speed. If your power is too light to maintain the speed regularly, do not take your governor off and run with a throttle; try the experiment of running your engine at a higher *regular* speed, put a larger diameter receiving pulley on your mandrel, and you will get better results both as to quality and capacity. This will be much better than the throttle plan, even if your speed does fall below that given in table; the regularity is the most desirable point to look after. Following is a table of speeds.

SPEED OF SAWS RUNNING 10,000 FEET PER MINUTE ON THE RIM.

72	in.,	530	revolutions	per	minute.	36	in.,	1,080	revolutions	per	minute.
68	"	560	"	"	"	32	"	1,225	"	"	"
64	"	600	"	"	"	28	"	1,400	"	"	"
60	"	640	"	"	"	24	"	1,630	"	"	"
56	"	700	"	"	"	20	"	1,960	"	"	"
52	"	750	"	"	"	16	"	2,450	"	"	"
48	"	815	"	"	"	12	"	3 260	"	"	"
44	"	890	"	"	"	10	"	3 920	"	"	"
40	"	980	"	"	"	8	"	4,600	"	"	"

RULES FOR CALCULATING SPEEDS, Etc.

PROBLEM 1. The diameter of driving and driven pulleys and the speed of driver being given, find the speed of driven.

RULE. Multiply the diameter of driver by its number of revolutions, and divide the product by the diameter of the driven; the quotient will be the number of revolutions of driven.

PROBLEM 2. The diameter and revolutions of driver and the revolutions being given, to find the diameter of the driven.

RULE. Multiply the revolutions of driven by its diameter and divide the product by the revolutions of the driver; the quotient will be the diameter of driver.

SPEED INDICATOR.

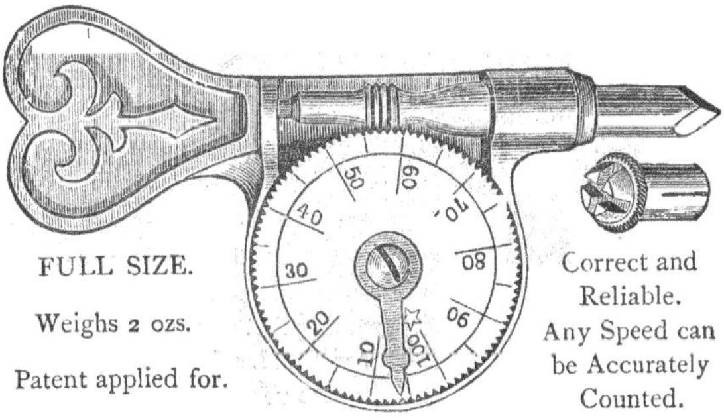

FULL SIZE.

Weighs 2 ozs.

Patent applied for.

Correct and Reliable. Any Speed can be Accurately Counted.

Sent by mail, prepaid, for $1 50, with full instructions for using. Every mill-man should have one.

This instrument is very useful for accurately giving the speed of any machine or shaft when in motion. It is very important that the exact speed of saws be given, and with the Speed Indicator there can be no mistake.

CIRCULAR SAWS OR DISCS FOR CUTTING HOT OR COLD IRON, OR STEEL.

For fast cutting of cold steel or iron, friction-discs, run at high speed, are used; for hot steel and iron, saws having teeth varying from 3/8 to 7/8 inches in space, are used, the angle of teeth being equally divided from a line drawn from points of teeth to centre of saw, and are run at a speed somewhat lower than friction-discs. These saws and discs are made of mild steel manufactured expressly for the purpose. The teeth of the saws should be frequently sharpened and kept square on both edges, and the flanging on the discs, which becomes crystalized and ragged, should be taken off, or the great strain put on the discs and saws by the high motion and heavy pressure of work will surely break or tear the metal apart.

To give the best results, these saws and discs must be run at a high and uniform rate of speed. Hot saws should be run at about the rate of twenty thousand feet per minute (rim motion). Discs for cutting cold iron or steel should be run about twenty-four thousand feet per minute (rim motion).

It is highly important that the mandrel and collars upon which these saws or discs are run should be amply heavy, large and true, and so secured in boxes to the frame work or housing as to do away with all undue vibration.

In cases where the motion of these saws is reduced from temporary causes the feed should be reduced proportionately, and when it has been reduced to any great extent, work should be suspended altogether until proper speed can be regained and maintained. The work in no case should be forced suddenly upon nor crowded on the saw.

The flanging which takes place upon the rim of discs should be removed before edges of same become ragged, or cracks in the plate will be the result.

Hot iron saws should be sharpened frequently and not run until teeth are half worn off, as this will cause them to jam in their work, and is liable to break them.

SETTING THE HUSK OR SAW FRAME.

It is quite as important that the saw frame should be firmly secured to its place by through bolts as that it should be level and solid, as the vibration and strain are of such a nature that the frame would quickly change its position unless *very* firmly secured, and the slightest change would make a vast difference in the running of the saw and necessitate relining. In putting in the husk stringers, use well seasoned wood and put them down in such a manner that they cannot possibly change their position, then find the position of the husk on the stringers and fasten down securely with through bolts.

LINING THE SAW WITH THE CARRIAGE.

In the ordinary method of grinding circular saws, it was difficult to get the surfaces of a saw true near the centre, and in consequence, even when screwed between collars which are true, the saw would not hang properly on the mandrel. This difficulty has been overcome by our improved principle for grinding saws at the centre, insuring perfectly true surfaces and bearings for collars. In many cases, however, the saw is strained by imperfectly turned collars, and also by the saw not slipping freely up to the fast collar, the stem of the mandrel being a trifle large near the collar, and the saw being forced to its place by the nut, is made full on the log side. To prevent this, it is only necessary to file the centre holes to a nice fit, so that the saw can be pushed to its place by the hands.

The amount of lead required for circular saws varies. The least amount that will keep the saw in the cut, prevent it heating at the centre, and from scratching the lumber when running back, is best. If the lead into the cut is too much, the saw will heat on the rim; if the lead out of the cut is too much, the saw will heat at centre. We therefore give the least amount that is generally used, which is one-eighth of an inch to twenty feet.

From the various methods used for lining a saw with the carriage, we give what we think will be most easily understood: First, see that the mandrel is set perfectly level, so that the saw hangs plumb and true when screwed between the collars, and is flat on the log side. Draw a line running ten feet each way from centre and parallel with the saw; fasten a stick to the head-block, so that it comes up to the line at end in front of saw; run the carriage forward the twenty feet, and stick should be one-eighth of an inch further from the line at the back end.

A very good method is to have a tapering board, with a hole in the wide end to fit on the stem of the mandrel. Screw the board tightly between the collars; drive a screw through the narrow end for a point; measure the distance from the head-block to the point; then turn the mandrel half way over by pulling on the pulley, so as not to strain or disturb the board between the collars; run the head block up to the reversed side and set the mandrel to the required lead.

All end play must be taken out of the mandrel when lining a saw to the carriage, and the track must be laid solid, level and true, so that the carriage will run straight and smooth

A very common but erroneous way of trying lead is by stretching the line from head-block to head-block and then measuring from line to saw; this is no test unless saw and collars are absolutely *true;* the guide pins must not touch the saw, and the head-blocks must be set precisely the same distance from saw line. Unless everything is true you will most certainly be misled by this method.

COLLARS FOR SAWS.

To have a perfect running saw it is indispensable to have the collars perfect and well fitting; but, as a general thing, there is nothing more defective about a mill, and any deviation from perfection in them is multiplied as many times as the saw is larger than the collars. They should fit exactly.

Test the saw with a straight-edge, and if it is found correct, place it on the mandrel, tighten up the collars by hand, slowly revolve the saw, and, if found true, all right so far; now tighten up the collars with a wrench, test again with straight edge and see if the position of the blade has been altered; revolve saw as before, observing whether it runs true; if not, the fault is sure to lie in the collars, and you will be likely to ruin your saw and certainly get no good out of your mill until the defects are remedied.

For large saws we prefer collars that have a perfect bearing of about one-half an inch on the outer rim, the other part clear, as they hold tighter than a solid, flat faced collar, because they are more apt to come fair against the saw. Examine the collars carefully to see if they are true, for it often occurs in turning a pair of collars the tool springs from many causes, some of which are through pins, or hard and soft places in the iron, tool being too light, and from many other causes. In order to be certain that none of these causes shall affect the saw, our advice is, in all cases, before the mandrel is taken out of the lathe, a fine file of just the proper spring be held against the face of the collars; this will take off any of the proud or uneven points that may have been left on them.

We are now finishing all of our circular saws by a new and improved process, which insures each side of the saw plate being perfectly true, by this invaluable process, every particle of unevenness is removed; the saw never requires packing (providing the collars are true), and all the annoyance and trouble which has hitherto surrounded and perplexed the sawyer in this particular are forever removed.

CHISEL POINT SAWS.

We make four sizes of teeth and holders for the chisel-point saws which are numbered respectively, 1 and 2 Oregon Chisel Point and 3 and 4 Chisel Point.

The No. 1 Oregon Chisel Point being the largest, is best adapted to very large timber on account of having large throat or saw-dust chamber.

The No 2 Oregon also has large chambers, but is smaller than No. 1. A greater number of saw teeth can be put in the saw blade than of No. 1.

The No. 3 is smaller than the No. 2, and is well adapted to any kind of sawing.

The No. 4 is used principally for edger saws, or where more teeth are required than can be put in of the No. 3 size.

All of these teeth are inserted by improved machinery on the most correct principles, the holders and points being grooved on the same curved lines, making a perfect fit in the sockets. Points and holders are exact duplicates, and, if ordered as directed, will suit any saw made by us for that size of holder and point for which the saw was fitted. The stops in front of sockets give the teeth a good solid bearing, and prevent the holder from turning out in case a point should be broken off in the cut.

When the edges on the inner circle of the holder wear round, they should be filed square, which will prevent dust from passing and unnecessary wear on the saw plate. When holders become thin by long use, new ones should be put in, and the result will be better lumber and less wear on the saw plate. The amount of power required to run the Chisel Point Saw is much less than that required by any other.

When putting in the points, be careful to have them well oiled to prevent cutting the plate; see that they have a good bearing before using the wrench; to do this, it may be necessary to tap the point on the top with a light hammer until the wrench can be used without tilting the point backwards or causing it to fly out. If the point holder is not brought up full against the front stop, place a piece of iron on the holder where it projects under the point, and drive it home by a few strokes with a hammer.

To prevent unnecessary wear on the sockets and the saw plate, the points should be dressed or filed without taking them out of the sockets; the points are not so hard that they cannot be sharpened by a good file which we make for the purpose, but the filing must be done on the front of the tooth, being careful not to leave a sharp nick under the point, which would cause it to break off and be likely to injure other parts of the saw.

No. 1.

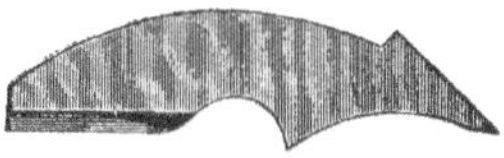

No. 2.

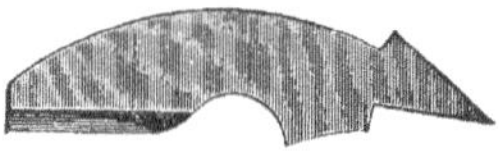

No. 3.

No. 1 shows point as it is when new; No. 2 one that has been properly filed until worn out; No. 3 bad filing, from which accidents may occur.

When ordering points, send a sample of the kind you are using.

The set should be carefully dressed on the sides by side-filing; if a swage is used, it must be done carefully with a *light* hammer, just sufficient to square the points.

When putting a point in place of one that has been broken off in the cut, be sure to have it dressed to size of those in the saw, so that it will not project further and be again broken off.

See that the guides are set so that they are below the sockets, for, if they come in contact with the holders, they will turn the teeth out.

In sending saws for repair, be sure to send the holders and a set of points, as it is necessary to have them in to properly hammer the saw; if not sent, instruct us if you wish us to put in a new set.

When sawing frozen timber use the worn points, select them in sets of one length, so that the saw will be round; swage the points a trifle for clearance; use no more set than is absolutely necessary; have the teeth widest at the extreme points, but not weak; taper the set nicely from point back by careful side dressing; do not allow the corners to become rounded or the saw will dodge out of the cut, more particularly in slabbing, as the points next to the log do the most cutting and soon become dull in sawing frozen timber.

In ordering these saws, be careful to give exact size of centre hole as any filing at this place would throw the saw out of round.

No. 1 OREGON CHISEL POINT.

Cut full size of tooth

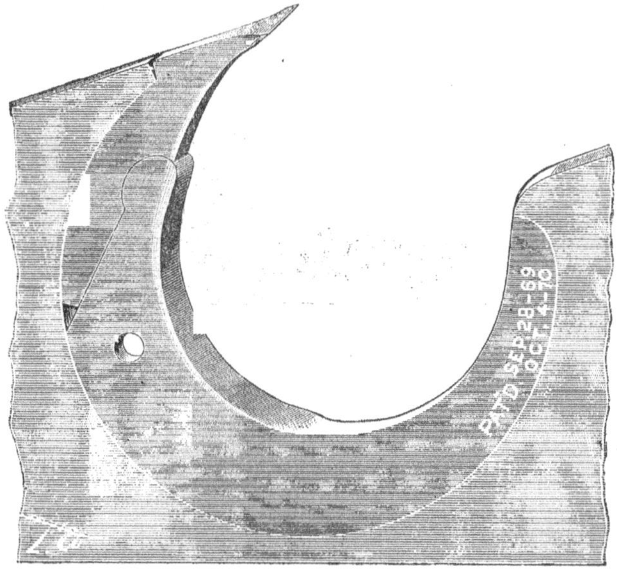

CHISEL POINT FILE.

Made 9 inches in length.

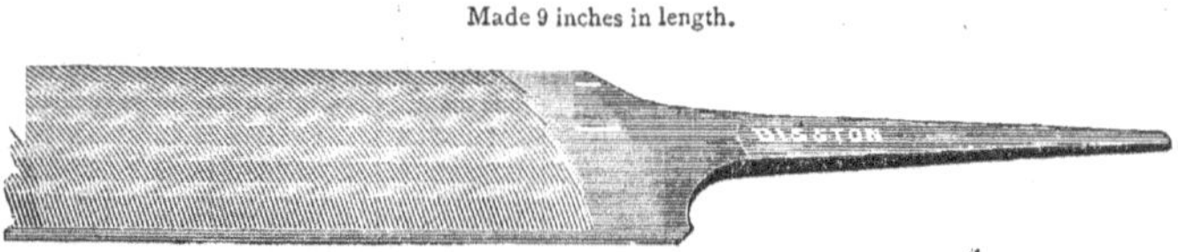

This file is made expressly for keeping the chisel points in the same perfect shape as they leave the factory; by its use the filing of sharp corners is prevented, from which points are liable to break off. See opposite page.

No. 2 OREGON CHISEL POINT.

Cut full size of tooth.

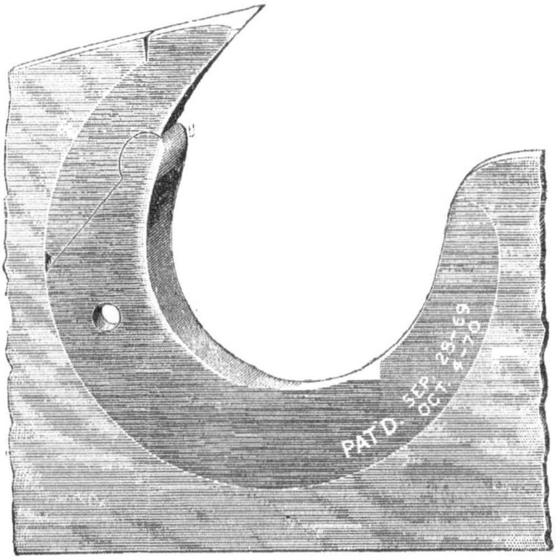

No. 3 CHISEL POINT.

Cut full size of tooth.

No. 4 CHISEL POINT.

Cut full size of tooth.

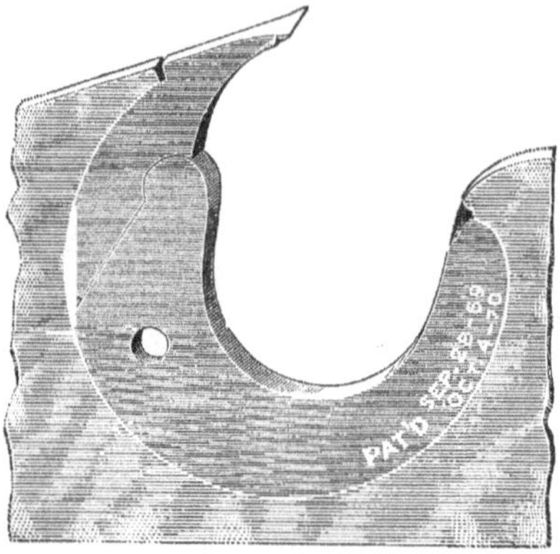

No. 10 TOOTH.

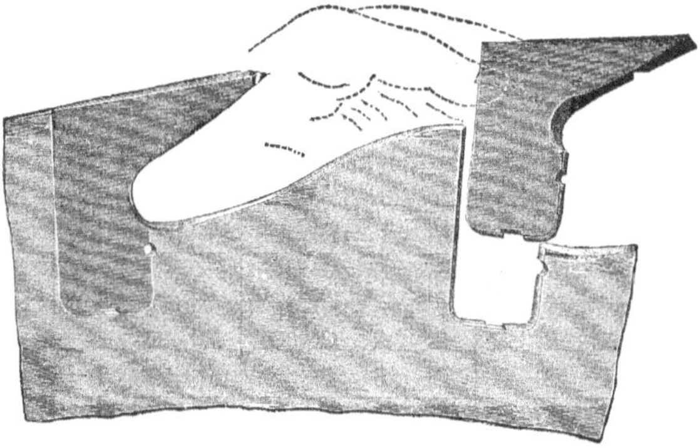

We call particular attention to the flanged tooth, by the use of which there is a great saving in power and prevention against the heating and springing of saws.

The flanges are so made in the saw plate and teeth as to cover the points of the teeth. The saw-dust formed by the cutting of the log is forced below the points of the teeth by the motion of the saw, and, without the flange, would pass the tooth and wedge between the sides of the saw and timber, thus heating the saw and causing unnecessary wear on the saw plate, requiring additional power to overcome the friction and loss of time in waiting for the saw to cool. From the above, the advantages of the flanged tooth are obvious.

Our saws are made with the greatest care by experienced workmen with improved machinery, and we use none but the best material in their manufacture. They are evenly tempered and ground by our patented process, which makes them perfectly true, dispenses with the liability of lumps or unevenness in the saw plate, whether of the same thickness throughout or beveled from eye to rim, and saves power, lumber and time.

All saws bearing the brand HENRY DISSTON & SONS, are fully warranted.

RE-SAWS.

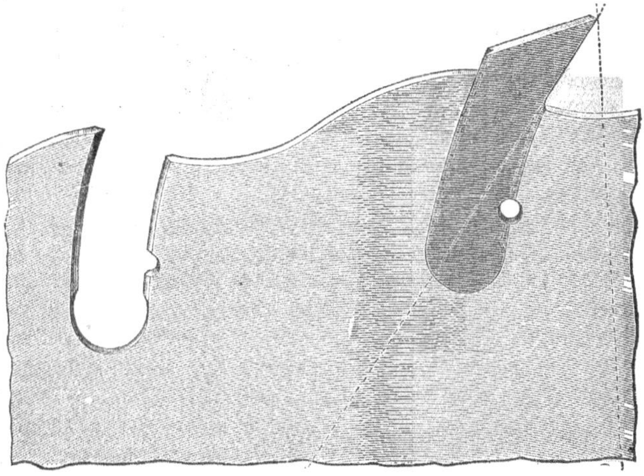

The difficulty occasioned by wearing down or reducing in diameter of shingle, veneer, segment, heading and re-saws, has created a more general demand for an inserted tooth saw of this class, and to supply this want, we are now making quite a large proportion of our segment, heading, shingle and re-saws with the improved re-saw inserted tooth, of which the above cut is a representation. The advantages claimed for this style of saw are numerous, the most important of which is that the original diameter of the saw is retained. This point will readily be seen by all practical operators and sawyers; for the saw must be the proper diameter and thickness at rim and centre to give the best results, for if the diameter is decreased, the periphery or cutting edge is brought closer to the heavy centre or flange of saw, not only cutting out a heavier kerf, but bringing an undue strain upon both saw and machine and causing the pieces being sawed to take a short, sharp spring-off, and in sawing short stuff where flanged saws are used, the flange or collar, by its close proximity to cutting edge of saw, splits a portion of piece from the bolt instead of sawing it, giving very unsatisfactory results both as to quality and quantity of work done. Therefore, if your saw is right at the start, by retaining original thickness and size, these difficulties are entirely obviated, and to do this, inserted tooth saws must be used, or the solid tooth must be frequently replaced.

This saw can be made in gauges from 14 to 17 at the rim. By replacing the teeth when worn out, practically you have a new saw at a trifling expense. We give the same warranty on these saws as on any other goods with our brand on. They are no experiment, having been used for a number of years with satisfaction and economy.

We have made such improvements in our manufacturing facilities and brought this tooth to such a state of perfection that, all things considered, they are the best, as well as the most economical saws used in the class of work for which they are intended.

KEEPING AND FITTING SAWS, Etc.

See that your saw slips up freely to fast collar and hangs straight and plumb when tightened up, that your mandrel is level, in proper line with your carriage, and that it fits in its boxes as neatly as possible without heating, for when the mandrel heats, by transmission, the saw will heat also and thus expand in the centre, which will make it work badly, injure, and perhaps ruin it. We do not warrant a saw to run on a mandrel that heats, although if we knew to what degree of heat you get your mandrel, we could make a saw that would admit of that much expansion, but a heating will *always* give more or less trouble. To get the best results from a mill, this heating must be overcome. See article on mandrels for circular saws.

Take up all end play or lateral motion in mandrel, as the grain of the wood will draw or push the mandrel endwise, no matter how well the saw is kept. See that the carriage track is level, straight, solid, and in proper line, also that rolls or trucks have no end play. Keep all gum or saw-dust off the tracks. Keep the saws sharp, round, and swaged or set enough for clearance; when everything is in proper shape $\frac{3}{32}$ of an inch equally divided is sufficient swage or set. Keep extreme point of tooth the widest. Do all the filing on the under or front side of tooth, filing square across the teeth, but never file square corners in saws of any kind, as this renders them liable to break. When you have occasion to swage or upset the teeth of your saw, file them all to a sharp point first; this will not only save time but save the saw, for the sharper the teeth the more easily will they upset or swage. Endeavor to keep the teeth in the shape they were when new, for if they lose any of the hook or rake or saw-dust chamber, it will not only consume more power but be harder to keep in order, as well as make inferior lumber. Keep your saw well balanced and your saw chambers well gummed out. By following instructions given with our No. 1 Gummer you will save your saw, time and files, and get much better results generally.

SHARPENING AND GUMMING WITH EMERY WHEELS.

Great care must be taken in sharpening or gumming saws with emery wheels. Always use a good, free-cutting wheel, and never put so much pressure on it nor crowd it so fast that you heat and blue the teeth, for when teeth are blued, glazed, or case-hardened by the emery wheel, they are apt to break or crumble when in the cut or the next time they are swaged.

When gumming it is best to go around the saw several times instead of finishing each tooth at one operation, for by going over the teeth several times, they are less liable to case-harden or blue, and gives a more uniform gullet. After gumming, it is advisable to file all around the saw, taking care to remove the fash or burr left on the edges and all the glazed or hard spots, as cracks will start from the little cuts made by the emery wheel which are not discernable at first, but by the use of the saw become deeper, and finally the teeth break out or the cracks extend into the body of the plate and the saw is ruined. Gumming and sharpening with the emery wheel will cause your saw to "let down" or lose its tension much quicker than by the use of the file or burr-gummer, as it heats and expands the rim of saw, putting it in the shape that is generally termed by mill-men "rim bound," which makes it appear loose and limber and causes it to run snakey in the cut. Many saws are condemned just from this cause and thrown aside as worn out, when by the proper work and hammering they can be made as good as new saws of the same size.

In sending us old saws for repairs, mark plainly on the case whom they are from, and write us full instructions as to what you wish done, and we will guarantee to put as good and durable tension in them as they had originally. We make and repair saws as well as alter from solid to inserted tooth.

PRICE-LIST OF KEYSTONE SOLID CORUNDUM AND EMERY WHEELS.

Diam. in Inches.	Rev. per Minute.	THICKNESS OF WHEELS, IN INCHES.													
		¼	½	¾	1	1¼	1½	1¾	2	2¼	2½	2¾	3	3½	4
3	4800	80	85	1 00	1 10	1 25	1 35	1 50	1 60	1 75	1 90	2 00	2 10	2 60	3 00
4	3600	95	1 10	1 40	1 65	1 95	2 20	2 50	2 75	3 00	3 30	3 60	3 85	4 60	5 00
5	3000	1 20	1 40	1 80	2 20	2 60	3 00	3 45	3 80	4 20	4 65	5 00	5 40	6 40	7 40
6	2400	1 45	1 75	2 50	3 05	3 70	4 35	5 00	5 65	6 30	6 95	7 60	8 25	9 55	11 00
7	2100		2 30	3 15	3 95	5 00	5 60	6 40	7 25	8 10	8 90	9 75	10 55	12 20	13 85
8	1800		2 60	3 60	4 50	5 70	6 40	7 60	8 30	9 60	10 20	11 60	12 10	14 00	16 00
9	1600	......	3 15	4 40	5 60	6 80	8 05	9 30	10 50	11 70	12 95	14 20	15 40	19 00	21 75
10	1500	......	3 70	5 15	6 60	8 10	9 50	11 00	12 40	14 25	15 35	17 25	18 20	21 10	24 00
12	1200	......	4 00	6 00	7 40	9 00	10 70	12 75	14 00	15 70	17 40	19 00	20 75	24 25	27 50
14	1050	......	6 20	8 70	10 70	13 20	15 20	17 80	19 70	21 95	24 20	26 45	28 70	33 00	37 00
16	900	......	......	11 00	13 70	16 00	19 40	22 25	25 00	27 05	30 80	33 65	36 50	43 00	48 75
18	825	......	......	13 75	17 00	21 75	24 50	28 75	32 00	35 75	39 50	43 25	47 00	54 50	62 00
20	750	......	......	....	20 00	25 00	29 00	35 50	38 00	42 50	47 00	51 50	56 00	66 00	75 00
22	675	......	......	....	....	29 90	35 50	41 12	46 75	52 00	58 00	63 50	69 25	80 50	91 75
24	600	......	......	....	....	....	43 00	50 00	57 00	64 00	71 00	73 00	85 00	99 00	113 00
30	500	......	......	....	....	....	....	....	91 00	102 00	114 00	125 00	136 00	159 00	185 00
36	400	......	......	....	....	....	....	....	127 00	....	157 00	....	188 00	219 00	250 00

THIN AND EXTRA THIN LARGE SAWS.

As we have said in the preceding pages, all saws and saw-mill machinery must be kept in the proper shape to obtain the best results; this is especially necessary in running thin saws, for while a thick or standard gauge saw will give very fair results where only medium skill in the management of saw and mill is used, a thin saw would fall far short of giving fair results under the same methods and management, for from reasons given below, a thin saw requires more skill and better appliances to give same results than a thicker one.

First, being lighter and thinner, it necessarily follows that they cannot reasonably be expected to stand as much crowding as a thicker saw.

Second, it is always necessary to have enough set in a saw to give good clearance, which means enough to prevent the log from rubbing on the body of saw.

In the ordinary gauges of larger circular saws, say 7, 8 and 9, used in the ordinary manner on the average feed and ordinary lumber, $\frac{3}{32}$ of an inch equally divided is about as little set as should be run except in hard woods and frozen timber, then less may be used. A thin saw requires just as much clearance as any other saw, consequently, in proportion to the thickness, the thin saw has the most strain to bear, hence, from these reasons alone the best skill and mill are required. We do not wish to convey the idea that we do not make and warrant thin saws, but simply wish our customers who contemplate putting in extra thin saws to take a plain logical view of the facts and figures relative to thick and thin saws. The difference in thickness between 8 gauge and 10 gauge is $\frac{1}{32}$ of an inch; the set for clearance of each being the same, $\frac{1}{32}$ is all it is possible to save in kerf, and between an 8 gauge and 11 gauge the difference is $\frac{1}{32}$ of an inch full, hence the saving in the instances above is very small—so small, in fact, that in nine cases out of ten it is offset by reduction in capacity or in poorly manufactured lumber.

As to saving in power, the difference in nineteen cases out of twenty is not in favor of the thinner saw, for, being so much lighter, it will deviate from its line much easier, and any deviation, even so slight as to be imperceptible in the length of the cut, will consume by friction all the power you save in difference of kerf.

These are plain facts which any man who knows the gauges can figure out for himself, and we advise every mill man to study the subject well before ordering extra thin saws. If his mill, skill of employes, and value of timber is such as to justify extra thin saws, then have them by all means, and we claim that our saws in workmanship, toughness, elasticity, and standing-up quality of steel are unequaled, whether thick, thin, or extra thin.

In ordering, please note that thin saws require more teeth than heavier ones to do the same sawing, as this equalizes the strain on the rim as well as prevents springing of the teeth.

Regularity of speed is desirable with all saws, but particularly so with thin ones, as they depend more than any other upon the velocity to hold them up to their work. In extra thin saws, one-sixth more speed than given in the table will be advantageous. See suggestions on keeping and fitting saws.

HINTS FOR COLD WEATHER.

Take the frost out of the saw with hot water before setting the teeth or putting the saw in motion, as steel is always brittle and easily broken when frosted. As many saws are broken in winter, owing to the great risk in sawing frozen timber, the greatest care should be taken not to put any undue strain on the metal by running a dull saw. Keep the points out full and square and do not allow them to become rounded, as the saw will dodge out of the cut, more particularly in slabbing, as the points next to the log do the most cutting and soon get dull in sawing knotty frozen timber. Use no more set than is absolutely necessary; have the teeth widest at the extreme points, but do not have them weak; taper the set nicely from point to back. Sharp corners should never be filed in the gullets as cracks are sure to start from such misuse of the saw, particularly in cold weather.

PITCH OF TEETH FOR HARD AND SOFT WOODS.

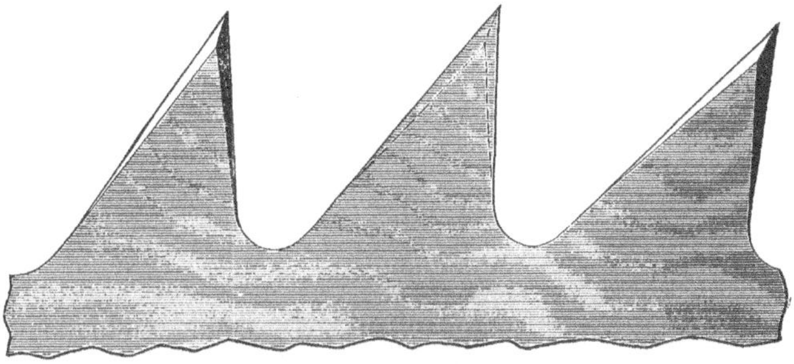

No. 1.

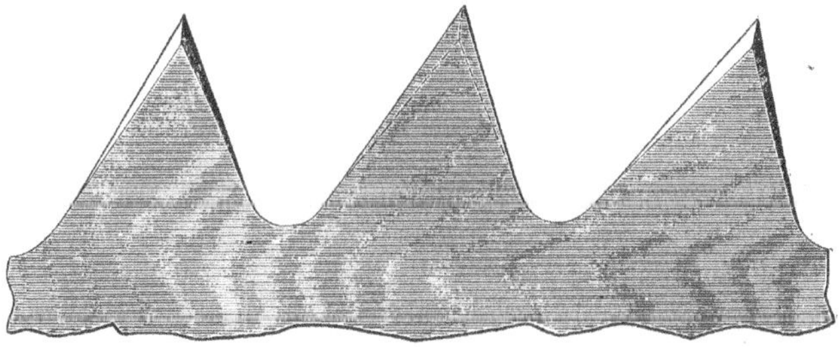

No. 2.

Cut No. 1 shows proper shape of tooth for cross-cutting soft wood. Cut No. 2 shows tooth best adapted to cutting hard wood.

TRAMMEL FOR CIRCULAR SAW TEETH.

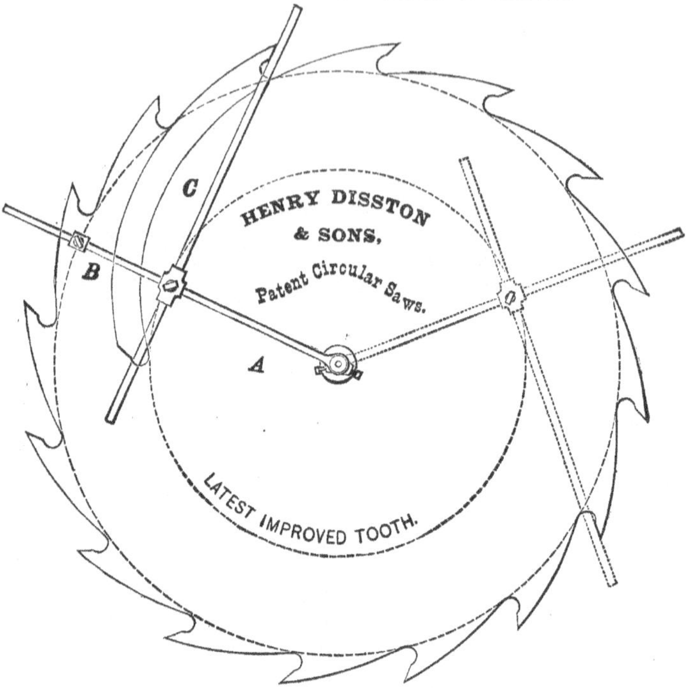

The above cut represents a device for laying out and keeping in order the teeth of circular saws. By its use the teeth can be kept in proper shape, regular in depth, and an equal amount of pitch given to the front of each tooth.

To rod *A* is attached chuck *B*, which holds a steel point for marking a circle for the bottom of the teeth. If all of the teeth are on this circle, they will be equal in depth. The strip of steel (*C*) can be set at any distance between the centre and the edge of the saw, and it will give the same pitch to the front of each tooth. The ordinary pitch is that which is obtained by placing the steel strip at a distance of three-fifths from the centre towards the edge of the saw-plate. There is a diversity of opinion concerning the proper pitch to be given to the fronts of teeth; knotty timber requires less than clear timber; with light power and light feed more can be used. The pitch can be increased by moving the steel strip nearer to the edge of the saw, but should the teeth become weakened, the backs or tops of the teeth should be strengthened, or they will either break or chatter in the work. Price, $2.50, net.

PATENT GULLET-TOOTH CIRCULAR SAW.

One of the most Valuable and Useful Improvements of the Age.

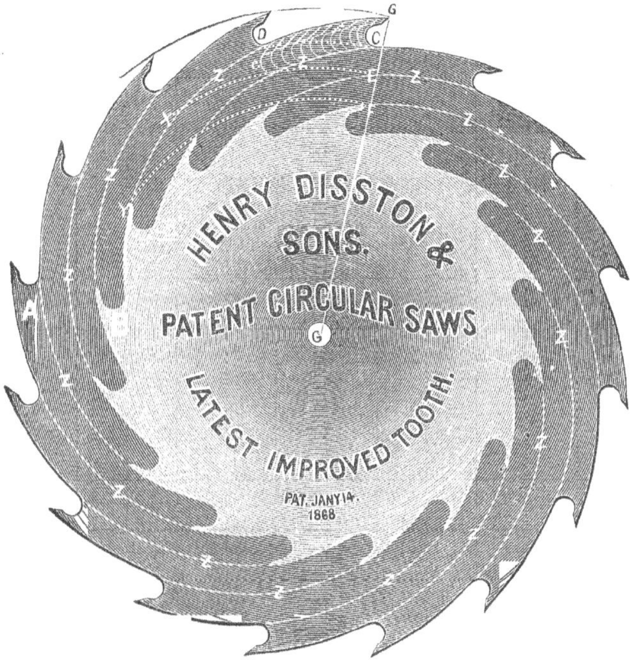

By reference to the above engraving, it will be observed that the back or point-line of each tooth is the continuation of the spiral lines *Z*, and the sharpening is mainly done by the reduction of the gullet or throat only. This is readily accomplished by the use of our patent gummers. (See pages 35 to 37.)

The course pursued by this cutter is spiral, and while it is in the act of reducing the front or throat of tooth *D*, it is prolonging the back or point-line of tooth *C*. The engraving represents a two-inch tooth or gullet. The saw *B* is the saw *A* worn down. When the saw has been reduced on centre line from *G* to *F*, it has been worn away but six inches, yet has presented a cutting surface on spiral line *Z* from *G* to *Y*, a distance of twenty-four inches. But this is only one of the advantages claimed for our patent gullet-tooth. The throat or gullet being chambered out on a half circle, forms a larger receptacle or chamber for dust, and thus a one-and-a-half-inch tooth of this pattern will keep a saw as free from choking as a two-inch tooth of the ordinary shape.

The saving of the saw-plate by the use of a smaller tooth is evident to the most casual observer. In wearing a fifty-four-inch saw down to a forty-two, a loss of twelve inches has been sustained in the diameter of the saw-plate, which is six sets of two-inch teeth, or eight sets of one-and-a-half-inch teeth, an advantage of two sets in favor of our new patent gullet-tooth, independent of the immense gain by gumming on spiral lines.

The chambering machines (represented on subsequent pages) are furnished with different sized cutters, and with them you work out your gullet or chamber for whatever sized tooth you may require; but the deeper your tooth the quicker your saw wears out. If you can do your work with a tooth one and one-fourth inches deep, your saw will last more than twice the time that it would if your tooth was three inches deep.

We find that some mills sawing hard and soft woods carry only about one-inch feed. In this case a very shallow tooth is required. A shallow tooth will cut the smoothest lumber; but, if you carry five-inch feed, as they do in some sections, you want a deeper chamber or more teeth certain, and the saw must wear down that much faster. What we wish to do is to give a tooth, where we can, to suit the feed and kind of work, for hard wood requires a different tooth from soft, in more ways than one.

A great saving in your saw-plate, time and files is effected by taking a good, deep, full cut, instead of a light, scraping one. As stated in a previous article, a tooth becomes dull on its face in proportion to the depth of cut taken at each revolution of the saw; for instance, with a thirty-second of an inch feed, it takes thirty-two teeth to cut one inch of lumber, whereas with one-eighth feed it takes only eight teeth to cut the same amount. In other words, the fibre or grain of the lumber has to be broken thirty-two times in one instance, and only eight times in the other; and when your tooth starts to break the fibre one-eighth of an inch in your log, it will do it with nearly as much ease and consume very little more power than if the cut was a thirty-second of an inch. Of course one tooth, in this example, becomes dull for one-eighth of an inch under the point, and the other only one thirty-second of an inch, but it consumes as much saw-plate, time and files to bring up one tooth as the other; it is, however, easy to overdo the thing; there is reason in this as well as in anything else. On tooth, Fig. 4, dotted line *B* shows where the point first wears; dotted line *C C C* shows how it should be filed back on the periphery line; but, too frequently, on account of the long surface to be filed, and the sharp corner in the gullet made as at *I* (which causes the teeth very frequently to break,

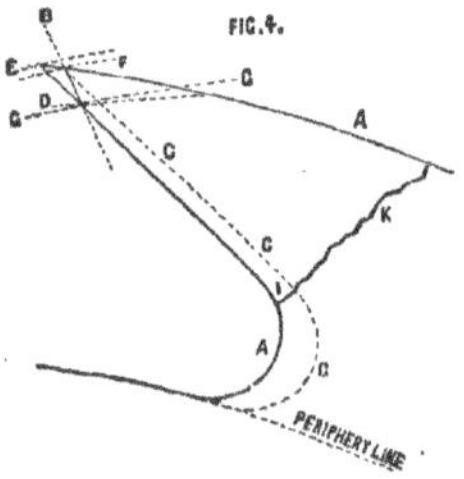

Filing back on the Periphery Line.

as shown at line *K*), they go to the top of the tooth, as represented by the dotted line *D*. You will see that by filing back on the dotted line *C C C* you have only diminished your saw from dotted line *E* to *F*, while by filing from the top you have diminished your saw, as shown by dotted lines, from *E* to *D*.

This shows that by filing on top you have wasted five times as much of your saw as by a proper filing. This, added to both sides of centre, shows you have actually diminished your saw ten times as much as you would have by proper filing. This difficulty is overcome by the introduction of our new tooth, as represented by cut, Fig. 5.

FIG. 5.

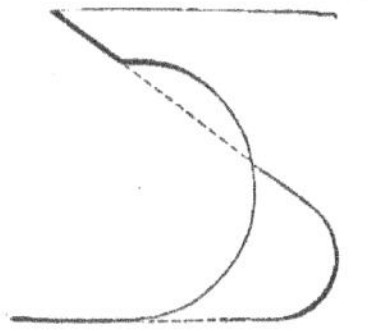

Showing Old and New Style Tooth.

You will see in Fig. 5 (same shape tooth as Fig. 4) by using the Gumming Machine (page 37), you have but a small surface left to file, which gives no excuse for filing on top of tooth.

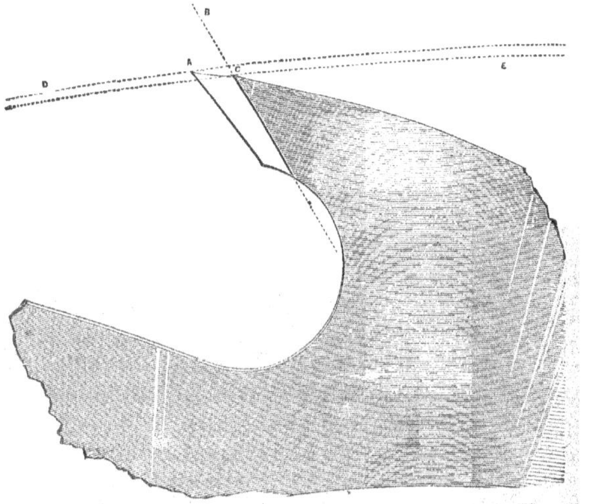

Engraving of Tooth after cutting 300,000 feet of Lumbe .

The above cut represents a section of our gullet-tooth saw (kept in order by Chambering Machine) after cutting 300,000 feet of hemlock lumber. Dotted line *D* and point *A* show the original diameter of the saw; dotted line *E* and point *C* show the saw after cutting the above amount of lumber, only reducing the diameter of saw about three-sixteenths of an inch, as can be plainly seen between dotted lines *D* and *E*. According to this, a fifty-inch saw will cut 9,600,000 feet, and only reduce the diameter of saw to forty-eight inches. You will thus perceive the great advantage derived by using our Patent Tooth and Gummer.

FIG. 6.

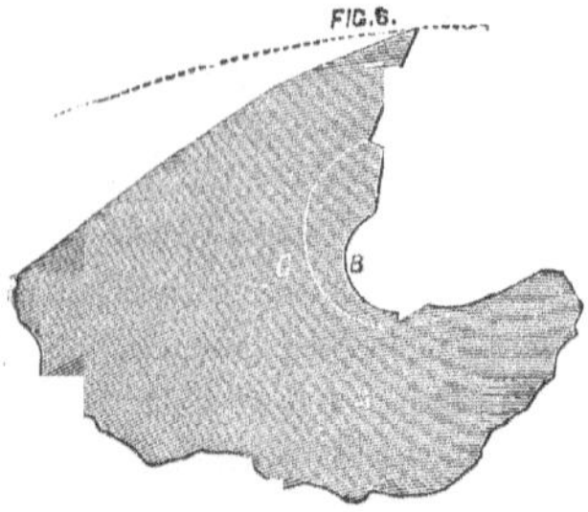

Bad Chambering.
(Reduced to one half natural size.)

The accompanying cut is a fac-simile of the condition of the teeth of a large circular saw sent to our factory to be ground. The parties had been using some gummer upon the saw, which actually did more harm than good; the reason we will explain: the line *B* shows where the cutter had been at work; but, being entirely too small, not even reaching the bottom of tooth, as it it originally was (leaving a ragged throat), and having so much depth to file at the top that the filer stopped before he reached the chamber, the consequence was, after a few filings there was a lump left in face of tooth, which so obstructed the circulation of saw-dust that the parties were compelled to send it to the factory to be gummed out. Dotted line *C* shows the condition the tooth would have been in had our chambering machines been used upon it.

Figs. 7 and 8 show, by periphery lines, the difference in the wear of the saw. We will here remark that it is of the greatest importance to file back on these periphery lines. You will see by this tooth the point on the face is very small. Well, the smaller it is the less filing it takes to keep it sharp. One stroke of the file on this point will effect more than ten strokes on the face of a tooth that you have to keep back from point to bottom of gullet; and when you have so little point to keep back, you will find it

FIG. 7.

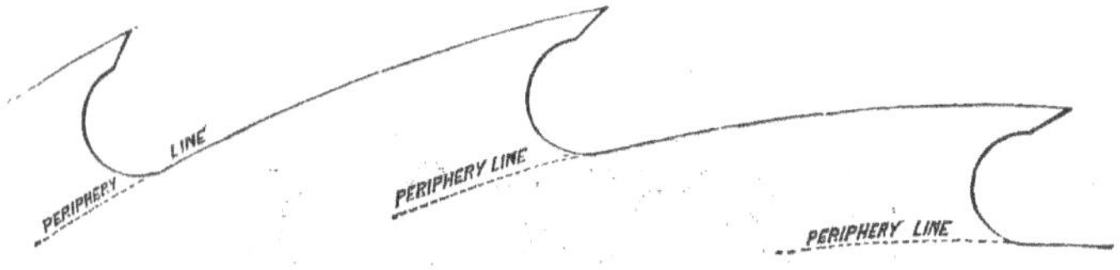

Teeth for Soft Wood.

easier to sharpen the saw from the face than to file from the top, and thereby you retain the diameter of your saw to a very great extent, as shown in Fig. 4, page 31; so much so, that you will not wear out one-third the saw-blade in one year that you do by the old process and old tooth.

If you could let us know what kind of lumber, and the speed and feed used, we could then give you the tooth you want, and, in many instances, save you waste of saw and the extra time it would require to keep a large tooth in order. For instance, for one-inch feed, we should not (where you use our gummer) give over one and a quarter depth; for a five-inch feed, not less than forty teeth, and depth to correspond; for a three-inch feed, we should give thirty-two teeth, and depth to correspond.

FIG. 8.

Teeth for Hard Wood.

The chambering machine ought to be put upon your saw every time that the file strikes the back of the chamber, as shown in Fig. 9, Tooth *A*.

FIG. 9.

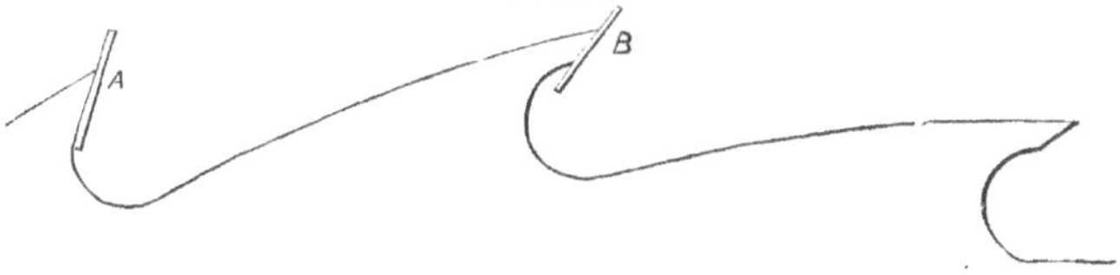

When your Tooth wants Chambering.

Tooth *B* has just been chambered, *A* wants chambering. By filing your saw back on this small point line, you wear your saw back on the periphery instead of (as is too much the case) on the centre line. Observe, if you file on the top of the saw, you go down towards the centre so much faster than when you keep back on the periphery line.

Fig. C.

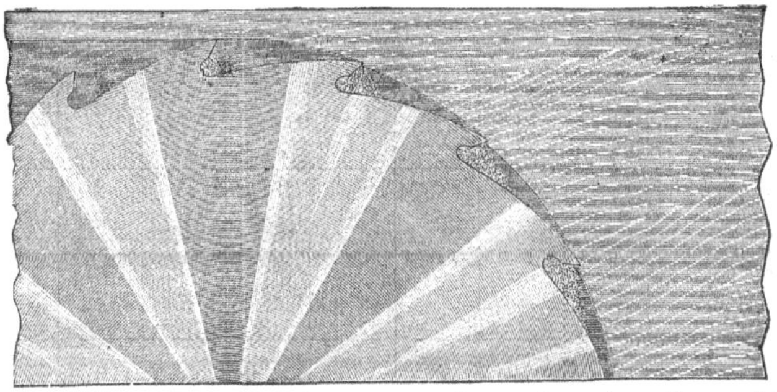

The above cut shows the No. 14 or gullet tooth saw in action, and the manner in which the dust is "pocketed" and carried out.

No. 2 PATENT SINGLE-GEARED GUMMER or CHAMBERING MACHINE FOR CIRCULAR SAWS,

38 INCHES OR SMALLER.

Adapted for either Circular or Long Saws. Cutter-Grinder and one Cutter each, ½, ⅝, and ¾ inch with each machine. Seven-eighths inch Cutter is the largest that can be used on this Gummer.

PRICE, $15.00.

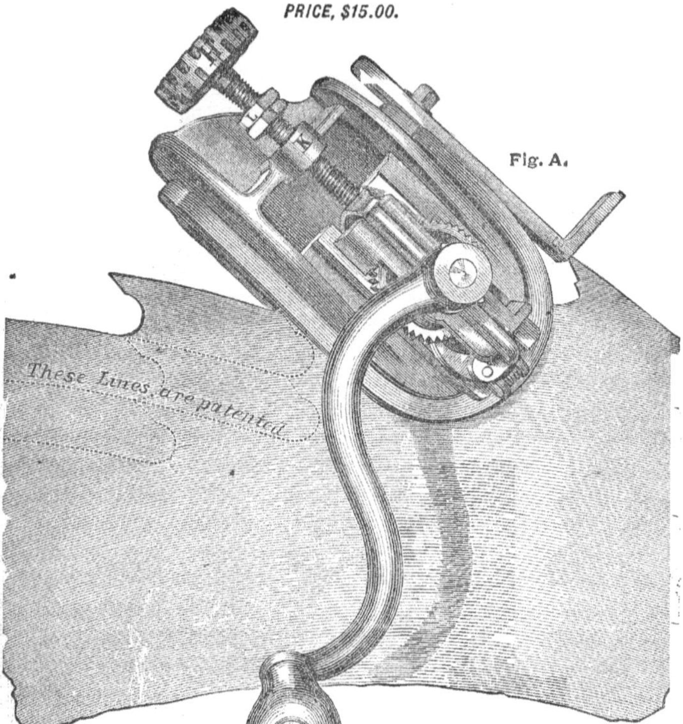

Fig. A.

Figs. A and B represent obverse views of the same machine, adapted for gumming either circular or straight saws.

Fig. A shows gummer in position on a circular saw. Put the gummer in proper position by means of screws *A* (Fig. B); run the screw *H* as far back as necessary; set the jam-nuts *L* to correspond with the desired depth of gullet. Gauge *F* must now be placed so as to strike the point of the tooth in front of it. Then proceed with the gumming until the jam-nuts *L* strike frame *K*. Remove gummer to next tooth, and repeat the operation until all the teeth are gummed. The jam-nuts must not be moved until the whole operation is completed.

No. 2 PATENT SINGLE-GEARED GUMMER or CHAMBERING MACHINE.

PRICE, $15.00.

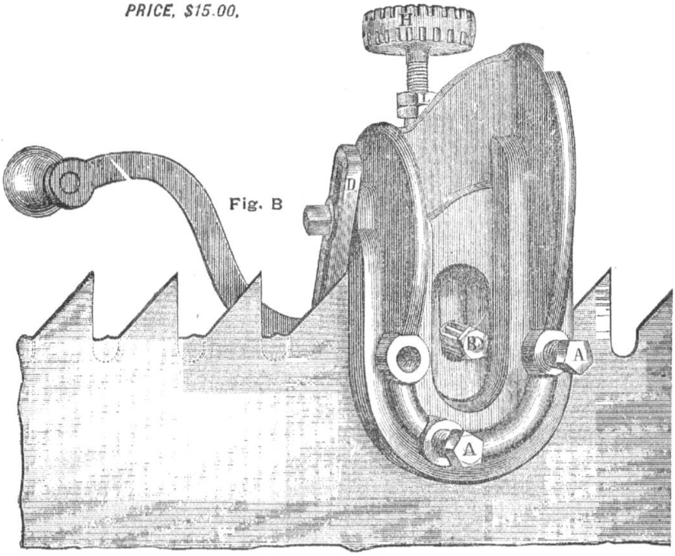

Fig. B shows the obverse side of Fig. A, with gummer in position on a straight saw. Place the cutter *B* in gullet of the tooth to be gummed. Secure the gummer in position by means of screws *A*; run the cutter back as far as necessary by means of screw *H*; set the jam-nuts *L* to correspond with the desired depth of gullet, and proceed to gum the first tooth feeding with screw *H* until jam-nuts strike frame *K* (Fig. A) When this tooth is sufficiently gummed, remove the machine to the next tooth, and proceed as before until the work is finished.

This machine has recently been much improved, an extra gauge *D* having been attached to the opposite side of the gummer, the two gauges *D* being used for the purpose specified. The cutter has also been furnished with a bearing on each side, which greatly adds to its strength and effectiveness.

GUMMER CUTTERS.

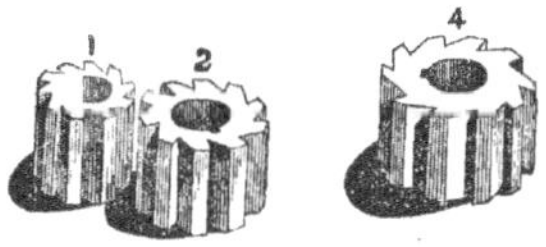

½ inch.	⅝ inch.	¾ inch.	⅞ inch.	1 inch.	1⅛ inch.	1¼ inch.	1⅜ inch.	1½ inch.
$0.40	.50	.60	.65	.65	.80	.90	1.05	1.20 each.

The 1 inch to 1½ inch cutters are for the No. 1 Gummer, and the ½ to ⅞ inch for the No. 2 Gummer.

No. 1 PATENT DOUBLE-GEARED SAW GUMMER, OR CHAMBERING MACHINE FOR CIRCULAR SAWS,

40 INCHES AND LARGER.

Cutter-Holder and One Cutter each, 1, 1¼, 1½ inch with each machine.

PRICE, $20.00.

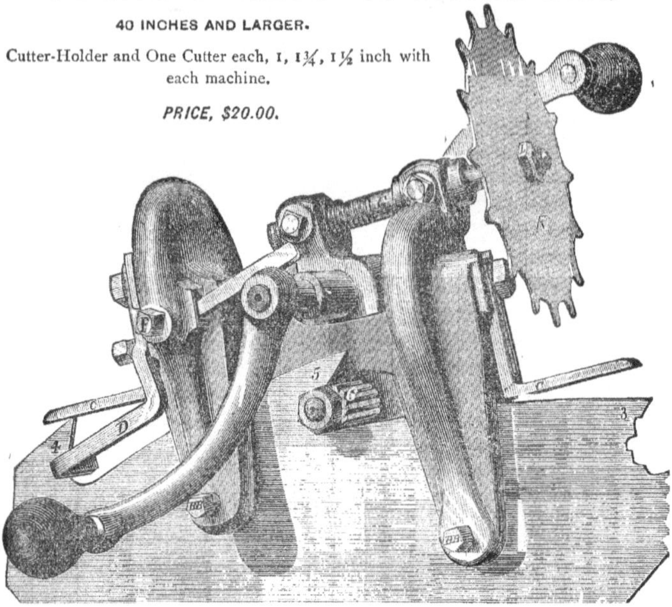

DIRECTIONS FOR USING.

Before using the gummer see that the oil holes are clear. A few drops of oil will be sufficient for from three to five hours' use. After using the gummer remove the chips or turnings that accumulate back of the cutter. If allowed to remain they will cause trouble by getting into the working parts of the machine. Run the cutter back by means of screw *G* as far as necessary. Then place the machine on the saw, with the cutter close up in the chamber of the tooth to be gummed.

If the teeth are regular and the same distance apart, start the cutter in any chamber; but if they are irregular, make them even by commencing in the smallest tooth. After gumming the saw a few times the teeth must become regular. *E* is a set screw to regulate the depth of gullet. Fasten the machine to the saw by means of the screws *BB*, and proceed to gum the first tooth, one of the points of the star being struck at each revolution by a projection on the handle. The cutter is steadily fed in until arrested by set screw *E*. Remove the machine to the next tooth towards you, after having run the cutter back, and proceed as before until the operation is complete. Should the gullet or chamber be worn smooth, and the cutter fail to bite, rough the gullet with a file. The cutter is so arranged as to slide on its axis, and when one portion becomes dull, by removing a washer from back to front, a new sharp-cutting surface will be presented, so continuing to change the washers until the whole face of the cutter becomes dull.

To take the cutter off the shaft, put the pin, hanging to the gummer, in the hole in the ratchet wheel *H*, to keep the shaft from turning while unscrewing the nut, which has a left-hand thread. The hand wheel on end of feed screw, outside of the star, is to allow the operator to feed easily and gently with the hand when starting in to cut rough gullets, until the cutter gets a bearing, when by tightening the jam-nut on opposite side of star, the machine is made self-feeding. The ratchet by which the cutter is moved, effectually prevents any back motion, which has hitherto been a serious objection.

This gummer is a most invaluable machine, and should be in the hands of every mill-man. It saves power, files and time, and is so simple in its mode of operation that any one of ordinary intelligence can be taught to use it. We pronounce this the best gummer ever manufactured.

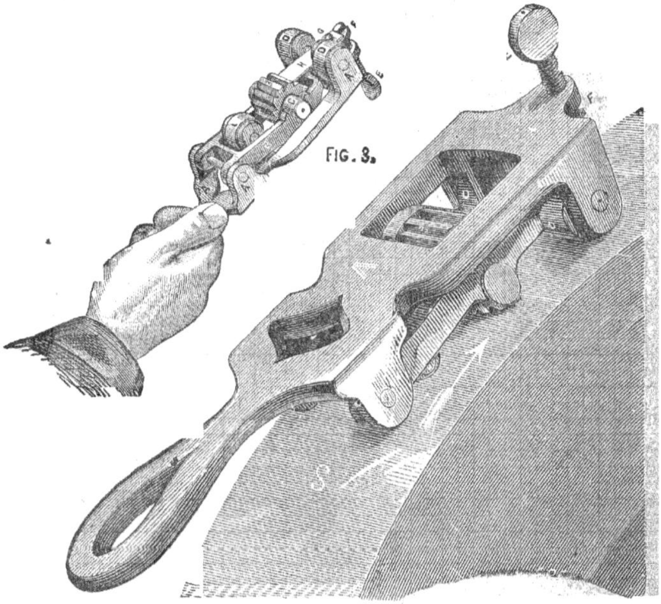

Device for Holding the Cutter of Chambering Machine in Position during the process of Sharpening.

A is the main frame; *D*, the rollers; *B*, the adjustable frame; 3, the cutter; *C*, the cutter shaft; *E F*, the screw for regulating the adjustable frame *B*; *G*, the screw for tightening spring *H*, which holds the cutter 3 in proper position while grinding; *S* is the grindstone. The stone should have a perfectly straight face and run through in the direction of the arrow. In using this machine care should be taken to hold it in line with stone. Screw the adjustable frame (which can be adjusted to any pitch) down till the cutter touches the stone; then see if the cutter is in its proper position; if not, it can be adjusted by spring *H*, by moving the spring either backward or forward.

No. 1.

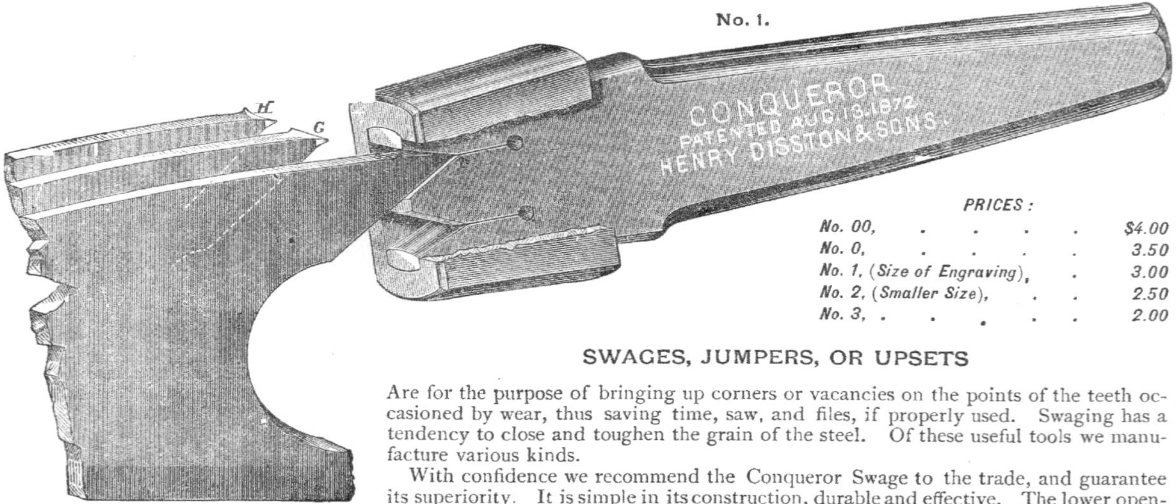

PRICES:

No. 00, ...	$4.00
No. 0, ...	3.50
No. 1, (*Size of Engraving*), .	3.00
No. 2, (*Smaller Size*), . .	2.50
No. 3, ...	2.00

SWAGES, JUMPERS, OR UPSETS

Are for the purpose of bringing up corners or vacancies on the points of the teeth occasioned by wear, thus saving time, saw, and files, if properly used. Swaging has a tendency to close and toughen the grain of the steel. Of these useful tools we manufacture various kinds.

With confidence we recommend the Conqueror Swage to the trade, and guarantee its superiority. It is simple in its construction, durable and effective. The lower opening, being rounding on the bottom, takes its bearing on the centre of the tooth, spreading and shaping it as shown in section of tooth *H*. The upper opening (in which section of tooth is inserted) is used for squaring-up, and leaves the tooth as shown at *G*.

The bottom of the openings in the Conqueror being slotted, it is impossible to blunt or injure the fine-cutting edge of the tooth, which is frequently done by other swages. One of the principal drawbacks in the manufacture of solid swages has been the difficulty experienced in properly hardening them at the Bottom of the Opening, which portion of the swage does all the work, and requires to be hardened in the most perfect manner. The Conqueror is hardened before the sleeve is driven on, and the hardening composition passes freely through the slots at the bottom of the openings, thus insuring an even and correct hardening. (See illustrations on next page).

SWAGES,

JUMPERS, or UPSETS.

Cuts full Size.

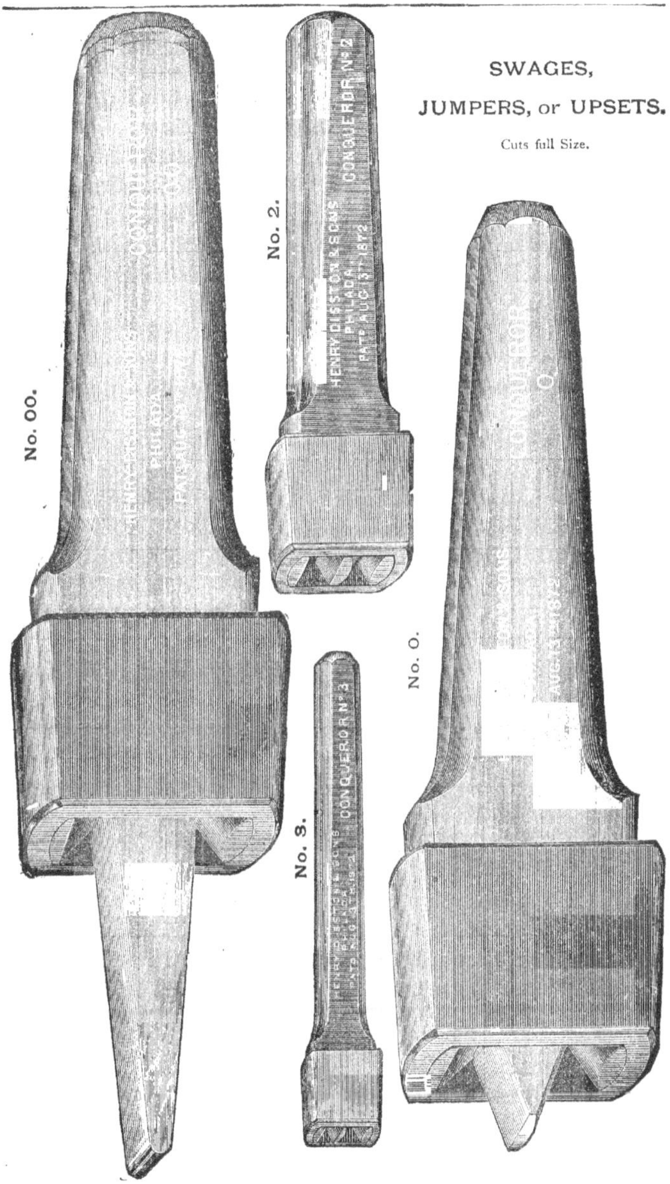

SIDE-FILE.

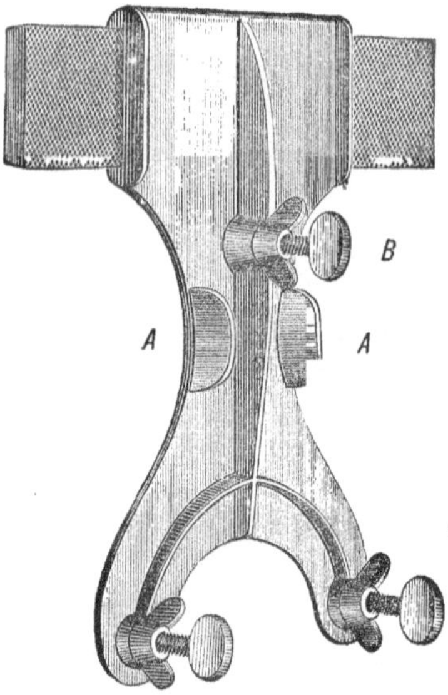

PRICE, $1.50.

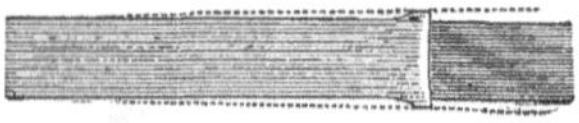

Double Spread Tooth.

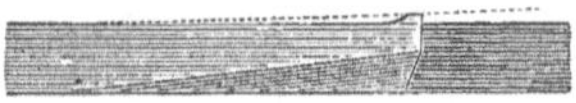

The Gridley Tooth.

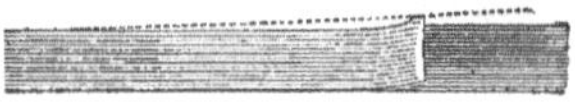

Spring and Spread Tooth.

The above cuts show the different sets for saw teeth.

The dotted lines show the undercut.

It is impossible to so set a saw that some of the teeth will not overhang, or are not bent over more than others. The slightest difference in the temper or variation in shape causes the teeth to set mild or rigid; this defect in the set, no matter how slight, must produce rough lumber. The difficulty can be overcome by the use of our side-file, an instrument invaluable to every sawyer or lumberman. It is *used for the purpose of regulating saw-teeth after they have been set*, and can be adjusted to any width of set required. It removes the extreme points and brings every tooth in perfect line on a firm foundation and true cutting-edge.

Every point is brought up to its work, leaving no vacant corners; no weak, flimsy, or extreme points, such as are left by the ordinary way of setting and sharpening. By the use of this tool, the set of every tooth is made even, and a saw thus regulated will run longer without sharpening. The file must be so adjusted by means of the set screws as to conform to the width of set desired. The jam-nuts are for the purpose of securing the set screws in the desired position. When the side-file has been properly adjusted it must be held in position by means of the clips *A*, against the saw blade, the points of the set screws *B* only touching the blade. Each tooth in succession must be filed until the set of the tooth conforms to the gauge of the set screws. Thus, all uneven or overhanging corners will be removed.

Shortly after the introduction of this invaluable tool, a customer sent to us for the machine which we had to help the saw to "*plane lumber*." We had some little difficulty to find out what he meant, but at length discovered that he wanted the "side-file." The immense demand for this useful little tool, and the testimonials we are constantly receiving, fully prove its value. When ordering, state plainly whether holder is wanted or only file for holder, or the side-file complete.

GAUGE BY WHICH TO FILE AND REGULATE THE SHAPE OF SAW-TEETH.

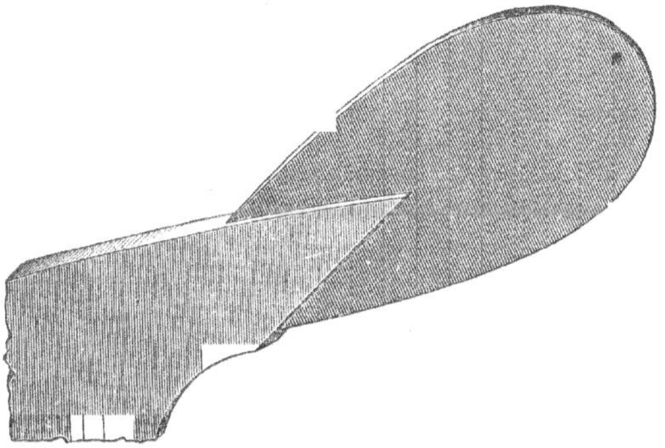

A tooth cannot be jumped or upset to advantage unless filed sharply and to the proper shape. To do this without a gauge requires a large amount of practice and experience. Taking for granted the back of the tooth is in good shape, the jumping must be done from the *under side;* this gives the proper rake and saves the saw. After the teeth are swaged, a few touches with the ordinary file and side-file complete the work. This useful little tool will be furnished gratuitously on application.

THE GRIDLEY TOOTH.

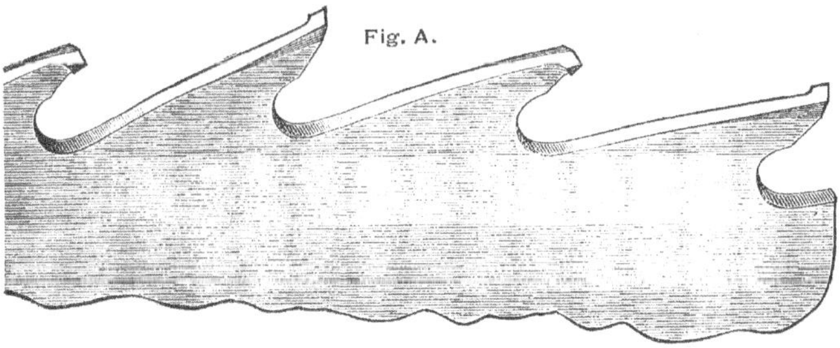

The attention of our readers is especially called to Figs. A and B. They are representations of the Gridley tooth, spring and spread set combined. Fig. A represents a portion of a saw in proper order for work, and Fig. B

Fig. B.

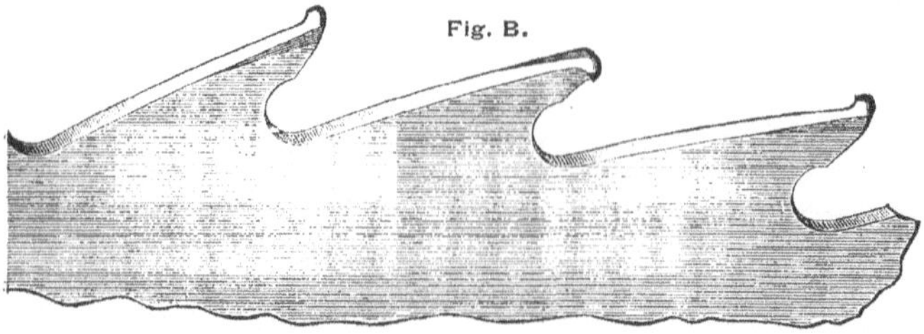

shows how the teeth become worn and rounded by use. Where the power is light and the sawing tough, this tooth, kept in proper order, will accomplish wonders. *The cutting-points must be kept up square and full,* as shown in Fig. A; for, when they become dull and rounded, as shown in Fig. B, it requires more power to run the saw, and makes inferior lumber.

In sharpening this or any other tooth, the filing should be done almost exclusively on the *under side;* the top should be filed sufficiently to keep the tooth in proper shape. The points of the teeth should *never* be filed to a keen, proud edge, but must be brought up by the use of the swage and a light hammer, and care should be taken to elevate rather than depress the point of the tooth during this process. After swaging, a few touches of the file will complete the work, and a great saving in saw-plate, files and labor will result. Thousands of dollars are annually wasted by the old method of filing and getting up the points of saw-teeth. Our saws are ground in the exact position in which they are used in the mill, and subjected to the same strain. Both sides being ground at the same time between two stones revolving rapidly, a true, even and uniform surface is the certain result, less set is required, and friction wholly avoided.

Many persons lose sight of the fact, that if the engine flags, everything about the mill flags in the same proportion, and the loss from this cause is immense.

HOW TO FILE CIRCULAR SAWS.

In a great many instances the persons who file circular saws pay no regard to the shape of the teeth; they get them into all kinds of shapes but the right one, as shown in engravings on next page.

The dotted lines on Fig. 2, page 44, represent the teeth in proper shape as they leave the factory; and the lines at *B, C* and *D* show the condition in which they are frequently sent to us for repair. In these teeth there is no chamber for the circulation of dust. At the same time this kind of filing wears away the saw five times as fast, and consumes three times as many files as would be required to do the work properly. But these are only a portion of the evil results. Teeth filed with sharp, square corners at the bottom frequently break, as shown at *A*, tooth *C.* It is lamentable to see this state of things when it can be so easily avoided.

FIG. 2

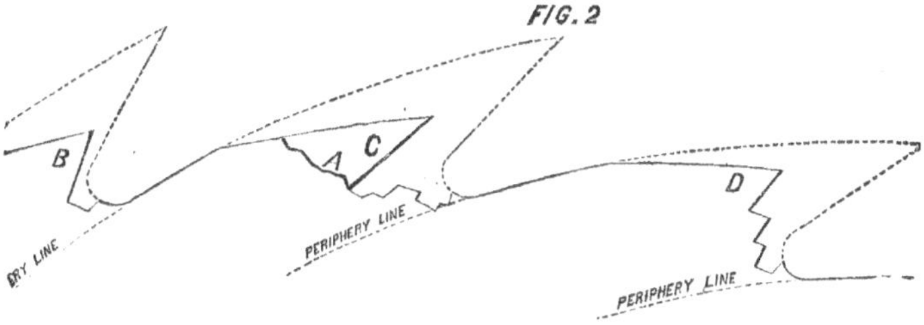

We have thousands of dollars' worth of saws returned for repairs, that break in cold weather by reason of these sharp corners. They are ruinous to the saw in many ways. If you will see to having your saw teeth kept in the shape they leave us, you will do away with this trouble and expense. The moment the teeth commence to get in bad shape, your saw begins to suffer in diameter, from the fact that the filer, wishing to get his points sharp too soon, files from the top instead of the face of the tooth. This does not help him one particle, but rather the reverse; and every stroke of the file on the top wears the saw away more than five strokes on the face or under side of the tooth.

JONES TOOTH.

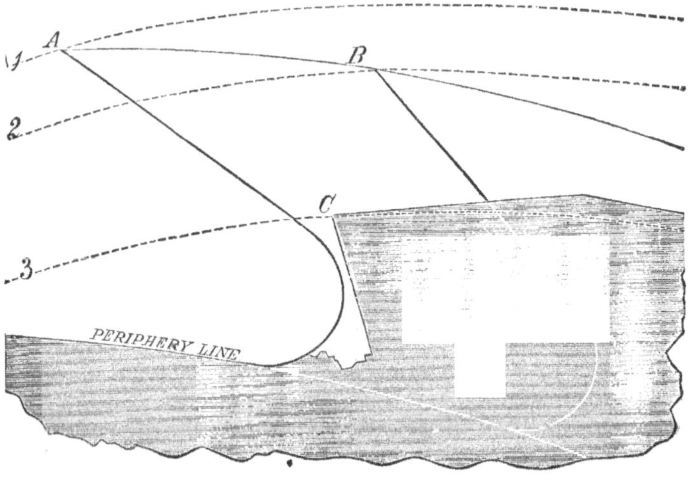

The above engraving of the Jones tooth will fully illustrate the evils of this pernicious and destructive practice.

This engraving represents a full-sized tooth, accurately traced from a saw sent to us for repair. The teeth of this saw had been filed from the top instead of the face. Dotted line 1 strikes the point of tooth at *A*, and shows the size or diameter of the saw when it left the factory. Dotted line 2 strikes the point of the tooth at *B*, and shows the size the saw would have been by filing back on periphery line, according to our directions. Dotted line 3 strikes the point of the Jones tooth at *C*, and shows how much he has reduced the saw by his fearful mode of filing. Take a pair of compasses and measure from *A* to *B*, and you will find exactly the same distance as from *A* to *C*; this, of course, presents the same amount of cutting surface in each instance, and yet in one case the saw has been filed down in size three times as much as the other.

Now, what has caused all this loss and trouble? Why, Mr. Jones has been filing from the top instead of from the face of his tooth. He has filed away and destroyed the useful portion of his saw, and retained that part which is of no earthly use to him, but is a positive injury. He has done three times as much filing as was necessary, and has consumed three times as many files as were required. He has left no chamber for circulation of dust; his tooth is higher at *D* than at *C*, thus instead of cutting his lumber with the point of his tooth, he scrapes it with the back. If his saw never commenced to heat before, it will do so now. Mr. Jones sent it back to us with the remark, "*It won't saw.*"

REVOLVING SAW-SET.

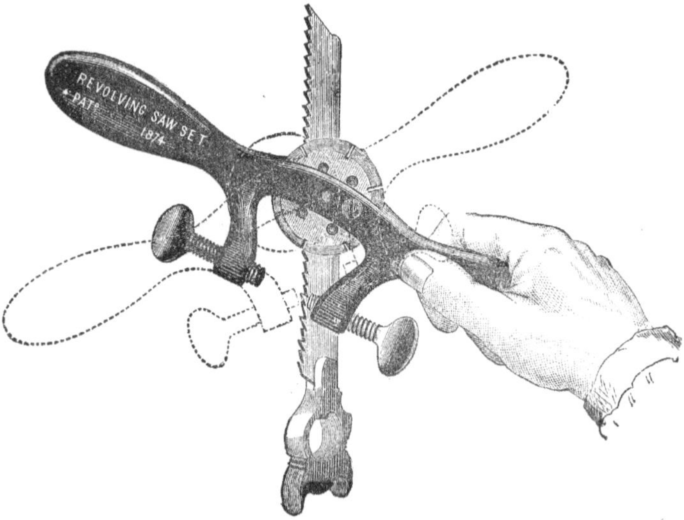

A good set for small circular saws, or saws with small teeth.

IMPROVED ADJUSTABLE SETTING STAKE FOR CIRCULAR SAWS.

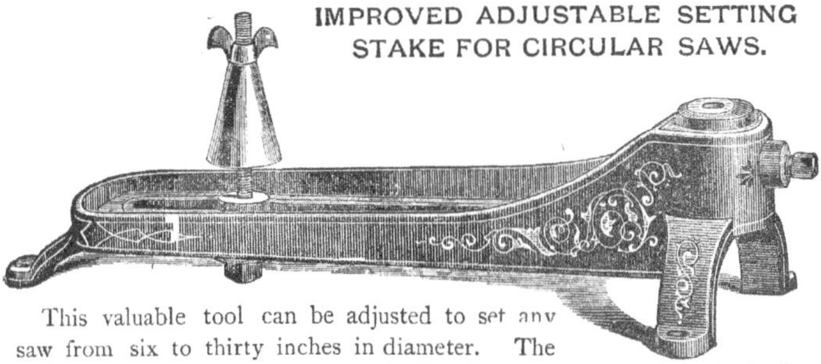

This valuable tool can be adjusted to set any saw from six to thirty inches in diameter. The cone *A* is moved in or out to suit the diameter of the saw, and raised or lowered as may be required. The movable anvil or *B* is made of hardened steel, and some portions of the face being beveled more than others, the operator can regulate the amount of set as desired.

SAMSON SAW-SET.

The Most Useful, Powerful and Desirable Made.

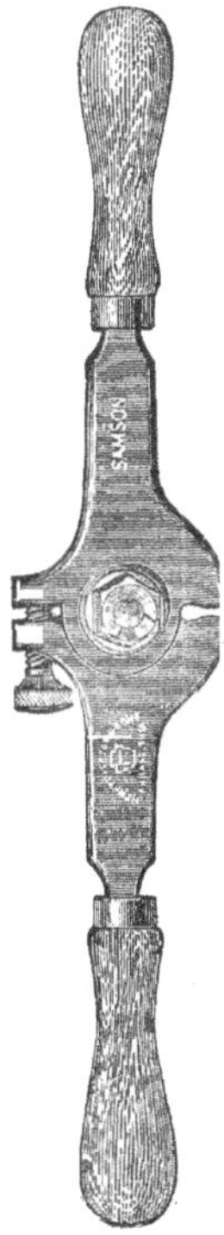

It is made of the best refined cast-steel, in two sections, and united in the centre by a bolt, which serves as an axis; thus it can be readily adjusted by means of the set-screw to suit the thickness or gauge of any saw.

THE BULLY BOY.

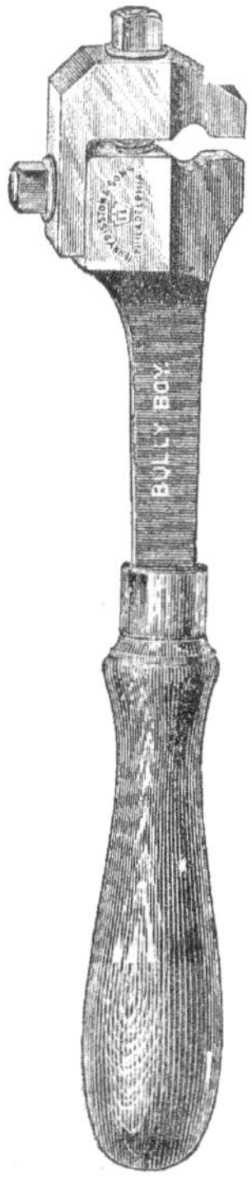

This is a first-rate saw-set, and can be relied on to give satasfaction.

CIRCULAR SAW MANDREL,

Of the Latest and Most Approved Style.

PULLEY IN CENTRE.

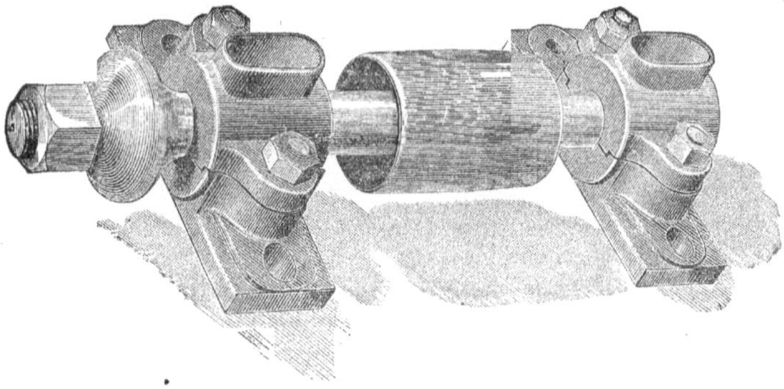

A mandrel should not be too light for the work to be done or it will spring, causing it to heat. See that the bearings are well proportioned and fitted. All bearings should be at least three times as long as diameter of mandrel: longer would be no detriment. The boxes should fit neatly enough to prevent lost motion, but not so tightly on the quarters as to cut off the supply of oil. One of the main causes of mandrels heating is want of proper lubrication. The cutting of channels from the front side of bottom half of boxes running down and under shaft to point of hardest bearing will be a great benefit in all cases; then use a good heavy body oil or lubricant. In some mills where three bearings are used on the mandrel, heating is caused by getting bearings out of line when shifting for lead or adjustment. Again, some mills are built with the collars for preventing end motion on the box nearest the saw; they should be on the other end, as the bearing nearest the saw has the most strain on it in addition to this at all times. Heating is often caused by having a too short and tight belt; where you have trouble with a heating journal and slipping belt, it would be advisable, as well as economy, to increase the diameter of the receiving pulley on mandrel, even at the sacrifice of some of your speed. Belts should be of good length, and in all cases should have the strain on the lower side and slack on the top; then when practicable, put a balanced tightener or stress pulley on the top, rigging it so that it will give as much lap of belt on the pulley as possible; this, with the balanced tightener, will take much strain off the mandrel, rendering it less liable to heat. A saw running badly from other causes, by undue crowding and straining will frequently cause a mandrel to heat that would otherwise run cool. See suggestions on keeping saw and mill in order.

We carry in stock mandrels with pulley on either end or in centre.

SHINGLE SAWS.

Left-Hand.

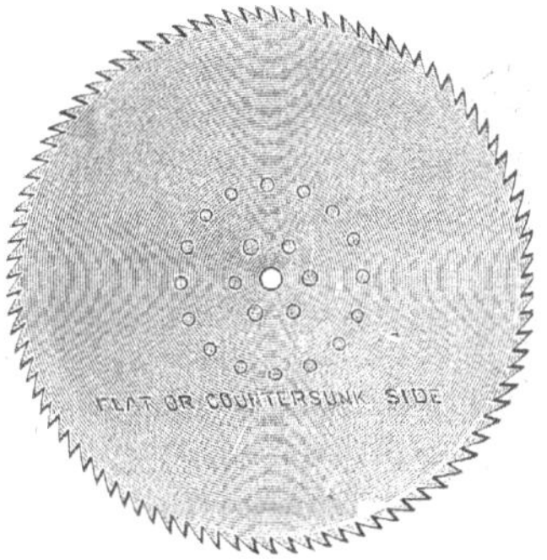

Right-Hand.

Fig. 1.

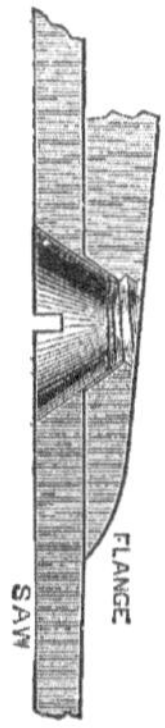

When ordering Shingle Saws, give the following directions plainly: diameter in inches; thickness or gauge at centre; thickness or gauge at rim; full sketch or pattern of holes, and sample of screw by which to drill and countersink. If you have a flange, send it to have holes drilled in saw to fit it. If you wish us to furnish the flange, send full and correct sketch of diameter, thickness, holes, etc. State whose make of machine the saw is to run upon, number of teeth, and be sure to give flat or countersunk side and the direction in which the teeth run (see engraving above).

Fig. 2.

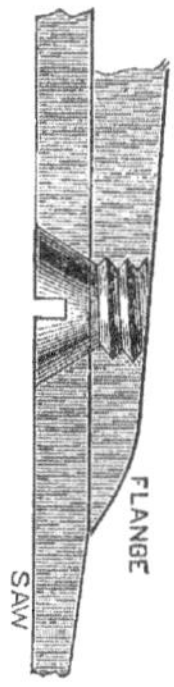

Particular attention is also called to the importance of using screws that are suitable for the thickness of the saw; we frequently receive screws as samples by which to drill and countersink, that have heads entirely too large, and which require the flange to be countersunk, (as shown in Fig. 1), thereby reducing the length of the thread and making it impossible to bind the saw firmly to it.

Fig. 2 shows the correct size the screw-heads should be, thus getting a good bearing for the screw-heads on countersink in saw and full thickness of flange for thread. In no case should screw-heads be deeper than thickness of saw.

VENEERING SAWS IN SEGMENTS.

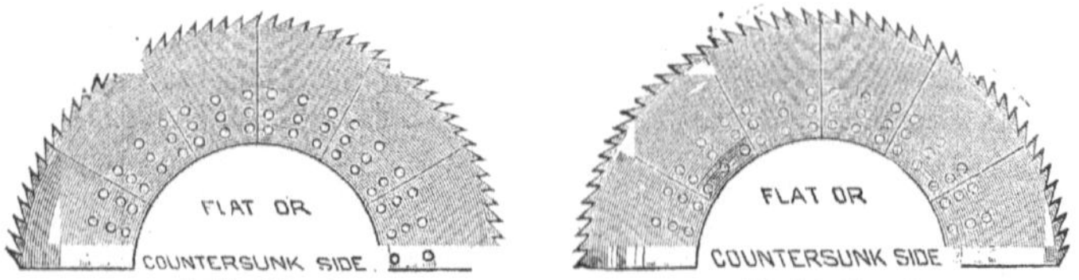

Left-Hand. Right-Hand.

When ordering segments, give gauge or thickness at butt and at edge, depth of bevel, diameter of saw that segments are to form, number of segments, depth of segments, number of teeth in each segment, sample of screw by which to drill and countersink, flat or countersunk side, and direction in which teeth run (see engraving).

In ordering for a flange that has been drilled, send a sheet iron or tin templet, or a correct tracing showing holes and other particulars; or one of the old segments, giving the depth they originally were.

CONCAVE SAWS.

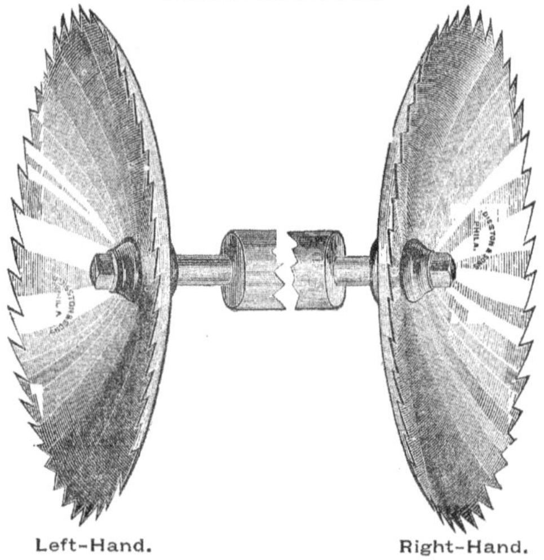
Left-Hand. Right-Hand.

The attention of the manufacturers of chair or wheelwright lumber, barrels, etc., is respectfully called to concave saws, of which we are manufacturing large numbers. They are dished and tempered by an entirely new and patented process, and guaranteed to be of superior quality in every respect. We furnish these saws considerably cheaper in consequence of our new mode of manufacture.

To keep concave saws in order, set both sides of the teeth alike; file the front of teeth square and bevel the back of each a trifle. Keep the same amount of rake on the fronts of all the teeth; do not run a dull saw, and keep the gullets round. If an emery wheel is used, be careful not to case-harden the saw. Smooth the teeth and gullets with a file after gumming.

CORK KNIVES.

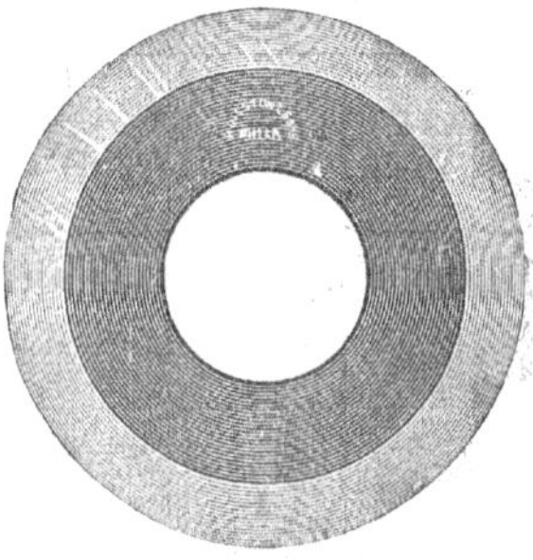

Cork Cutter.

Our stock for cork knives is made especially for the purpose and is the finest quality of edge-tool steel. This with our new process of tempering and grinding and the highest class of workmanship, enables us to turn out knives which for general superiority stand unequaled.

In ordering these knives give diameter, gauge, size of hole, whether to be beveled on both sides or only on one, and how deep bevel is to run. If knife is a large one and screws to plate or flange, send flange to us or an accurate tracing of holes, stating whether one or both sides are to be beveled, if only one side, state whether screw holes are to be countersunk on flat or beveled side. Circular and straight knives for cutting rubber, cork, etc., made to order.

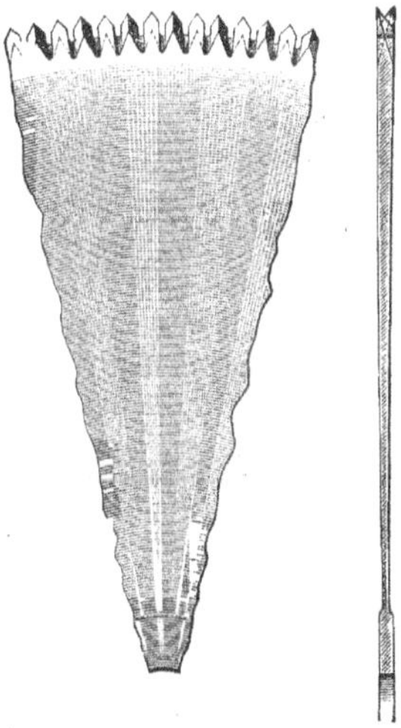

MITRE SAWS.

The above cut is a section of circular mitre saw and shows the manner in which they are ground, style of tooth, and how they should be filed. This saw is designed especially for cutting mitres and general small box work where clean, smooth and tight joints are necessary. It answers the purpose admirably, cuts as smooth as a planer, does true and rapid work, and is becoming more popular daily. This saw being ground thinner at centre, no set is required. In ordering saws of this kind, give size of mandrel-hole and collars of mandrel the saw is to be run on.

GROOVING SAWS.

These useful little tools are too well-known to require special mention. They are ground thinner at centre than at rim, so that little or no set is required or just sufficient to keep the extreme points of teeth perceptibly wider than body of tooth. We make them any gauge at edge or centre as may be wanted. In ordering grooving saws, state whether wanted straight or hollow ground, and if the latter, give size of collar.

GANG SAWS.

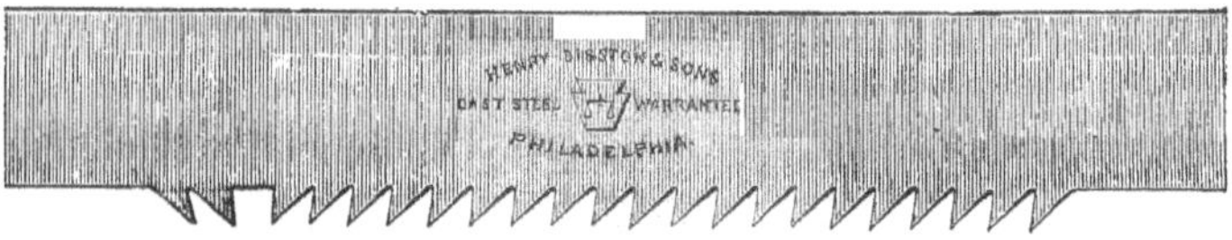

Our gang saws are made from steel which exactly suits their purpose, and they are evenly tempered and ground by our patented processes. We guarantee them to run more freely than those of any other make. In support of this assertion we cite the following instance:

Some time ago we visited a large gang-mill in this State which was run by two engines, each of one-hundred-and-fifty-horse power. The engines were flagging and the owners were considering the propriety of putting in more power. We induced them to try one gang of saws of our make, and the result was so convincing they immediately ordered three thousand dollars' worth of ours and abandoned the use of all others. They have since informed us that the change increased the capacity of their mill 33,000 feet per day, with power to spare. This saving of power was effected by furnishing them with saws made perfectly true and properly ground, thus overcoming all unnecessary friction. When filing gang, mill and mulay saws, care should be taken to have sufficient pitch to teeth to prevent "kicking back" and not too much to cause "chattering." The most desirable set to use is the square or swage set, and gullets should be round.

The reversed teeth in cut above represented will make a cleaner cut on the lower edge of the boards, preventing the tear ng or sprawling of the timber by the down cut.

BRAZING CLAMP.

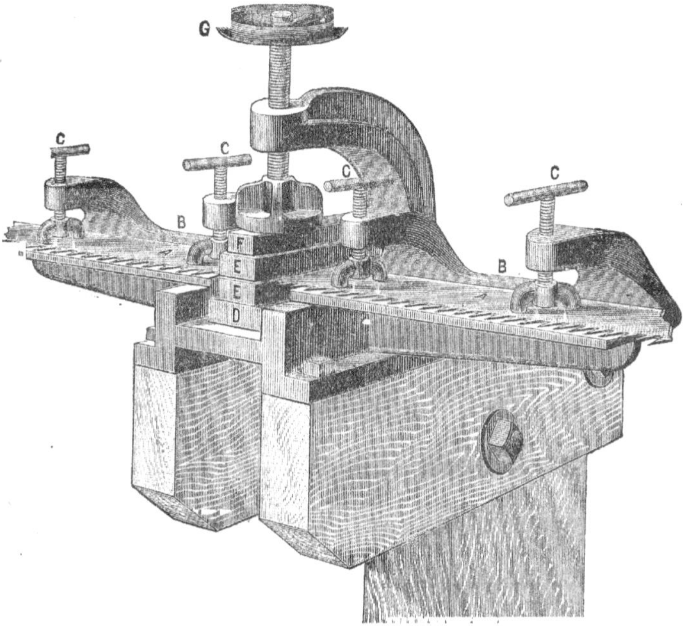

DIRECTIONS FOR JOINING OR BRAZING BAND SAWS.

The parts to be joined must be beveled on opposite sides to a width of three-quarters of an inch to a nice fit; the ends of bevels should be perfectly square and taper of bevel uniform througnout. Place the ends of saw on table with the back of blade against straight edge *B*, have the centre of lap over centre of irons *E* and *D*, then clamp securely by setting down screws *C C C C*. Clean the beveled parts with dilute muriatic acid; cover with a thin borax paste; cut a strip of silver solder same size as lap, clean this in the same manner as the parts to be joined, and place it between lap. Have iron *D* centrally adjusted; heat the irons *E E* to a bright red in a moderate fire, using charcoal, coke or hard coal. When the irons are at the proper heat, scrape all the scale from the sides to be applied to the saw; place them as shown in cut, centrally over and under the lap; place iron *F* on iron *E* and clamp firmly with screw *G*, after which, to allow expansion, slack up the screws *C C C C*; leave the saw in this position until cool enough to set the solder, tightening up occasionally on screw *G* as the irons contract by cooling, and when cool enough to handle, the joint can be cleaned, straightened and dressed to thickness of balance of blade.

HAMMERING BAND SAWS.

In hammering band saws they must be perfectly straight on the edges and open uniformly throughout the width of blade, dropping about one-sixteenth of an inch at the centre and tapering up to nothing to within one-half of an inch at both edges. A good saw, hammered in this manner and used on a good machine, will run well and give good results generally. Many saws are made that are unevenly tempered and hammered, which leaves long and short places in the edges or not of an even tension, and when subjected to strain are twisted and distorted, and finally break at the badly tempered places.

RUNNING BAND SAWS.

When running band saws, keep the correct pitch upon the front of the teeth so as to give a proper lead into the cut, which will prevent the saw from being pressed too heavily against the back guide, a continued heavy pressure against which will flange and crystallize the steel, from which cracks are sure to start. Band saws should be kept well sharpened, as many are broken from being run with dull teeth. Keep the gullets round. Select files with round edges for filing them. The saw must be uniform in width throughout, so that every tooth will perform its part of the work, for it can be readily seen how easy it would be to break the saw if the points of some of the teeth would project any considerable distance beyond the others.

WHY BAND SAWS BREAK.

In explaining the methods of adjusting and keeping band saws in order, we have shown many of the reasons why they break, but we give the following additional causes: Machines which have wheels entirely too small for the saw; starting or stopping the machine too suddenly; dirt clogging on the face of the wheels, making an uneven surface; any hard substance falling between the saw and the wheel; contraction of saw, pulling it apart, particularly in cold weather; the strain should be taken off when the machine is stopped. Saws, after gumming, should be filed, as they are very apt to break from the ragged edge put on by the gumming tool.

When we consider the number of times these saws run over the wheels and the strain put on them, it is only fair to ask of the user that they be kept in order as instructed in the preceding paragraphs.

DIRECTIONS FOR MAKING MURIATE OF ZINC.

Feed into muriatic acid small pieces of zinc until the mixture ceases to boil, after which dilute with an equal portion of rain or distilled water.

DIRECTIONS FOR PREPARING BORAX FOR BRAZING.

Roast the borax until all the moisture is driven off; pulverize and mix with distilled water to a thin paste.

DIRECTIONS FOR JOINING SMALL BAND SAWS.

The parts to be joined must be beveled to a nice fit. Secure the saw at both ends in clamps, as per cut. See that the edges are parallel, or a short

and a long edge will be the result, which will cause the saw to run badly and to break on the short edge when strained. Put on the filed parts a thin coat of borax paste. Cut a piece of very thin sheet silver solder of the same size as joint to be made, which place between the lap. Take a pair of tongs having suitably sized jaws for the joint and that have been heated to a bright red, sufficiently to melt the solder. Scrape all the scale off between the jaws with an old file; hold the joint with the hot tongs until the solder has thoroughly melted; remove the hot tongs carefully and follow up with another pair heated to show a dull red, which will set the solder and prevent the joint from being chilled too suddenly. The joint can then be dressed to thickness of the saw blade. It would be as well to have a pair of cold tongs to clamp the hot jaws firmly to the joint, as the hot iron must fit nicely over the whole width of the saw. In joining, do not make the lap longer than is absolutely necessary; one half-inch is sufficient for scroll saws, three-quarters of an inch for saws two to eight inches wide.

ANVIL, HAMMERS and STRAIGHT EDGES for HAMMERING SAWS.

The above cut represents the tools necessary for altering or adjusting the tension of saws. We make these tools with great care, and of steel best suited to the purpose.

SWAGE BAR AND HAMMER.

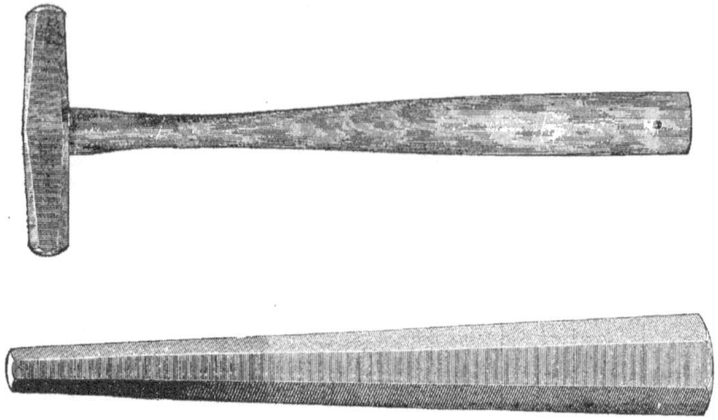

The above cut represents our swage bar and hammer for use on circular and gang saws. We make the hammers in two sizes; the bars of any shape, size or weight as desired.

GAUGE FOR REGULATING CLEANING-TEETH.

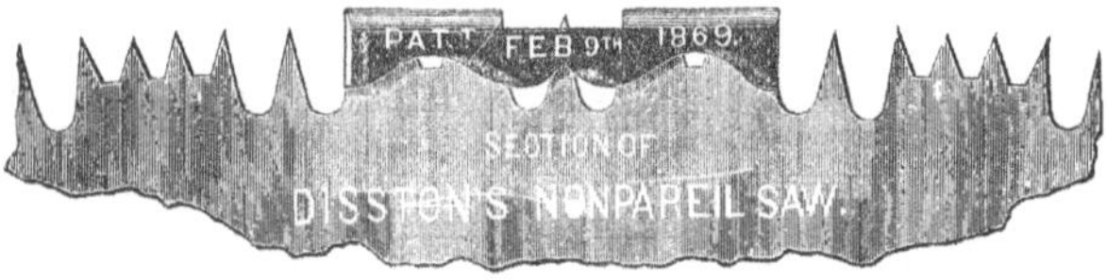

Showing the Gauge in Position for Filing the Cleaner-Tooth.

The cleaning-teeth of all saws should be somewhat shorter than the cutting teeth, and, although shortened, should be of uniform length throughout.

The inner edge of the gauge rests on the points of the cutting teeth, the cleaning teeth projecting through the opening in centre of gauge. Reduce the projecting points by means of a file, until arrested by the edges of the gauge, which is made of hardened steel. Thus tooth after tooth can be rapidly and correctly reduced to an even length by any unskilled operator.

THE GREAT AMERICAN.

Patented October 4th, 1870.

This saw has been subjected to the most severe tests, and is one of the BEST CROSS-CUT SAWS ever offered to the public. The most important advantages of this saw are as follows :

The outer teeth of each section are as sharp and effective cutting as the teeth of a rip saw, while the middle or regulating tooth determines the extent of the cut in proportion to the bevel of said tooth. The more the centre tooth is beveled the faster the saw cuts; whereas, if the centre tooth is filed square the saw takes less hold on your log, and requires less muscle to drive it. Thus the saw can be regulated to suit the strength of the parties working it.

With this saw there is no "*tearing of the wood, undue friction and drag*," which in many other improved cross-cut saws demand so much muscular exertion without a commensurate result.

There is no cross-cut saw in the market by which so much work can be done in ten hours, with so little exertion, as with the "*Great American Regulating Cross-Cut.*"

THE LUMBERMAN

is greatly preferred in some sections of the country, and can be easily kept in order if filed according to directions, while so many of the fast-cutting saws of the present day lose their shape.

In filing this saw, the round-edge mill-file should be used, and by pressing a little downward as well as sidewise the tooth is kept in the same shape it leaves the factory.

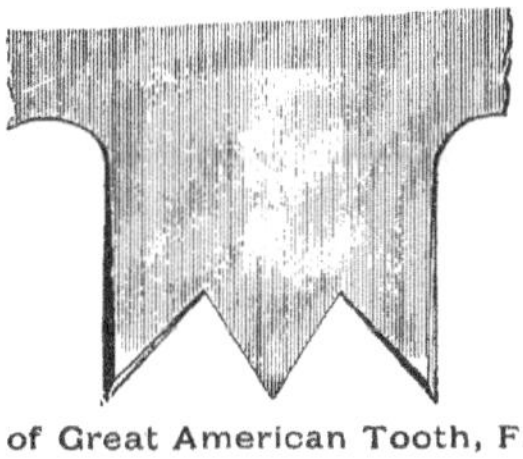

Section of Great American Tooth, Full Size.

ONE-MAN CROSS-CUT SAW, WITH SUPPLEMENTARY HANDLE.

Patented O tober 4th, 1870.

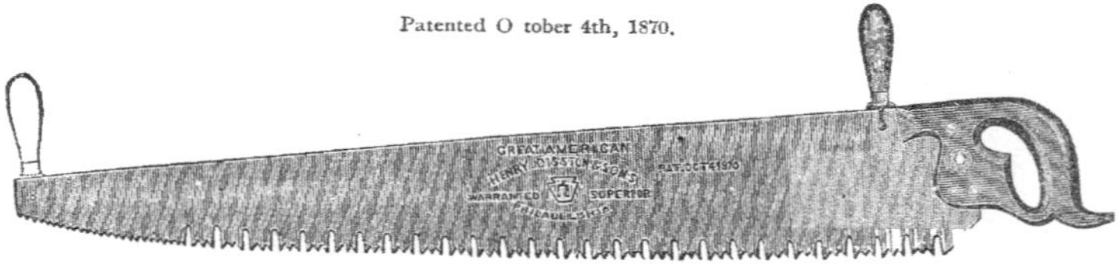

This engraving represents a cross-cut saw, especially adapted to the use of one man. The "Great American" one-man cross-cut saws are made and ground on the same principle as our No. 7 hand saws. We have lately improved the file for keeping this tooth in order, and it should be ordered with the saw.

It will be seen that the hands of the operator will be in an easy position. The handles and saw are bound together by strong iron plates and screws, a principle much needed in these saws.

Bridge-builders, mill-men, railroad and other contractors—in fact, all large establishments—will find this a very useful tool, and easily worked. For cutting off girders, joists, blocking, or heavy lumber of any kind, it is just what is required. As it is now becoming a general practice to saw lumber with one man instead of two (from the fact that one good man will at all times do more work than two poor ones), the saw will pay for itself in a few days, as the labor of one is saved. The above engraving illustrates the "Great American" tooth, but any kind of tooth will be subs'ituted to suit the various localities.

CHAMPION TOOTH No. 1.

TRIUMPH NARROW CROSS-CUT SAW, WITH HANDLES COMPLETE.

These saws have been made by us for many years, and their utility for cutting down trees can best be told by their use. They are not so liable to bind by kerf-closing.

Made with any style tooth desired.

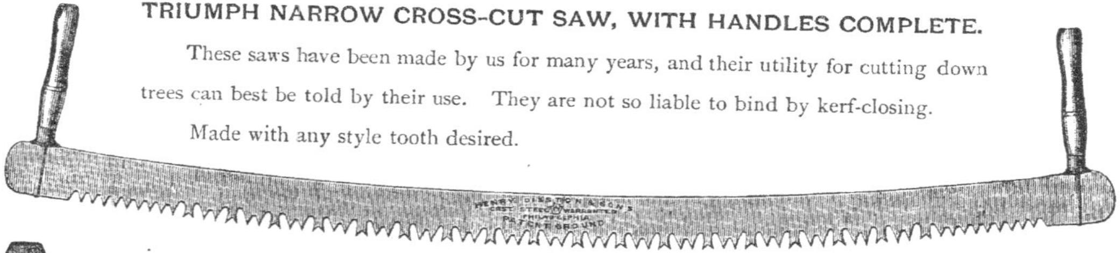

Reversible handles sent with these saws if so ordered, otherwise loop handles will be sent.

PATENT CROSS-CUT SAW HANDLES.

Patented May 9th, 1871.

Loop Handle.

Cross-Cut Saw Handle, Plain.

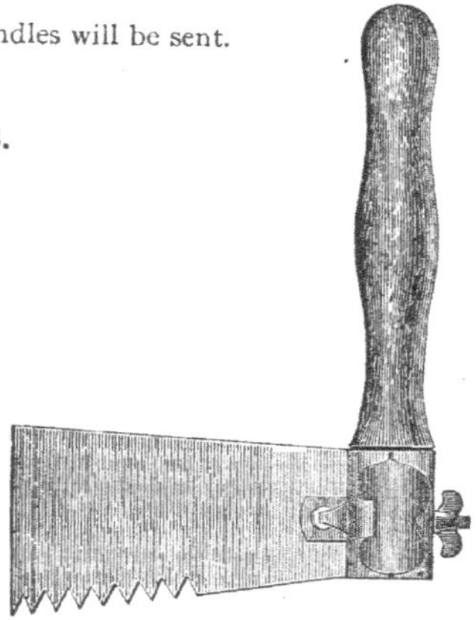

Climax Cross-Cut Handle.

FILES.

We use a great number of files in the manufacture of saws, employing over one hundred expert saw filers who file the various kinds of saws with a still greater variety of files. Each filer purchases the files he uses, and is allowed to use any brand of files from which he can get the best result so that we have the best practical means of knowing the proper and suitable conditions required in the manufacture of all the regular and special files for filing saws.

MILL SAW FILES.

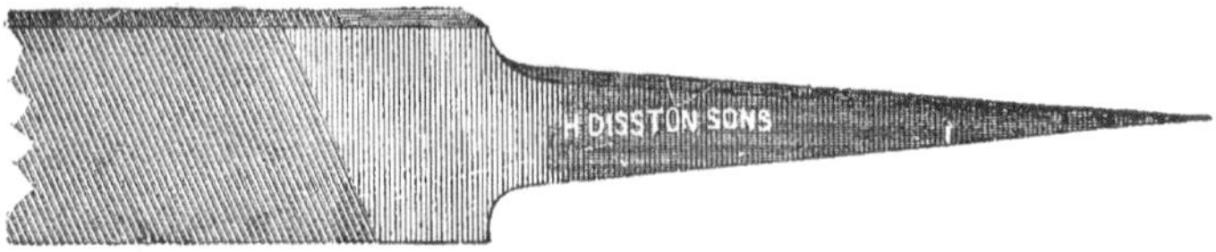

Our regular mill saw files are of the best crucible file steel made, and superior in hardness and sharpness to any files on the market. The teeth are so formed as to be most durable and to file keenly as well as smoothly.

TAPER SAW FILES.

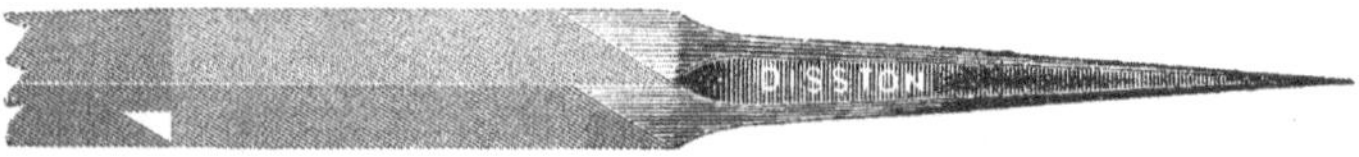

Our taper saw files in shape and quality are as perfect as the practical experience of the best saw filers can determine. The corners are carefully cut and made the proper thickness to be strong and sharp, and are in every respect made on the best known principles.

GREAT AMERICAN FILE.

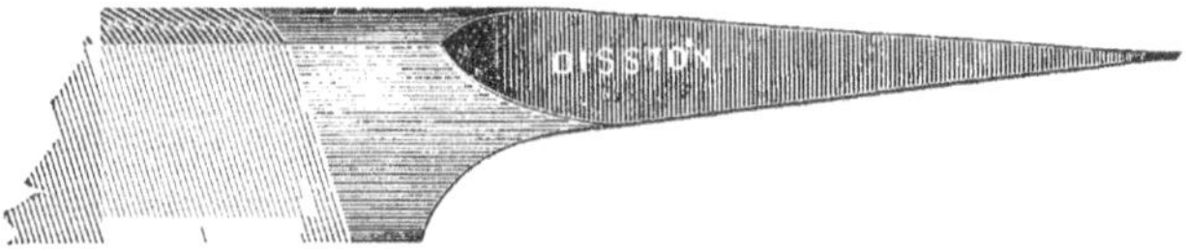

Our Great American cross-cut file was first designed and made by us for keeping in order the teeth of our Great American cross-cut saws. We have given special attention to their manufacture and can confidently guarantee them superior in every respect to any other files made for that purpose. It is also the most suitable file for the "Lumberman" and other cross-cut saws of that style.

NEW PROCESS OF SHARPENING FILES AFTER HARDENING.

The sharpening of file teeth after hardening is a valuable improvement in the manufacture of files, and makes those so manufactured (under this process) as far superior to any made under the old system as a sharpened saw is to a dull one.

All cutting tools, except files, are sharpened after having been hardened and tempered, and while it has been deemed essential that files should be treated in the same manner, it has been impossible to sharpen the teeth until this recently-discovered method makes it practicable to do so, thus insuring a sharpness that has long been desired, and which cannot be otherwise obtained.

New mill files sharpened by this process will not only do more work, but will make finer and sharper cutting edges, and will file a harder saw than the ordinary file. A trial of files made under this improved process will fully prove their superiority, and can be relied on to have a degree of excellence never before attained, and this without extra cost to the purchaser.

Rasps and files for brass and other metals are specially benefitted, and all packages containing such rasps and files will be labeled sharpened, and day and date when so done.

STAVE SAW FILES.

We make several styles and sizes to suit the particular kind of saw or the manner of filing. Every file is warranted. (See Price List.)

PATENT "GRIP" FILE HANDLE.

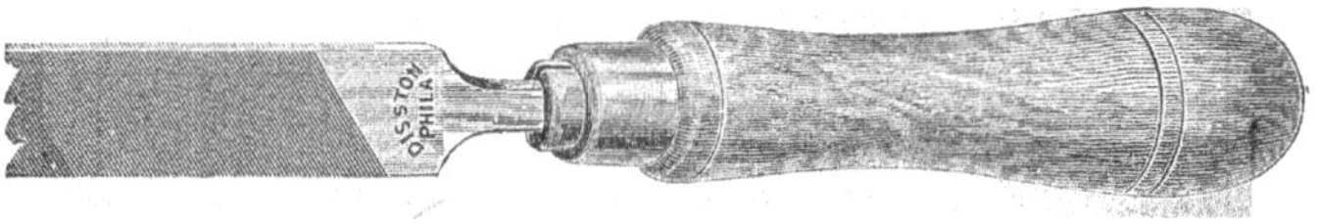

This file handle can be instantly attached or detached. The "grip" is hard and sharp and fastens into the soft tang of the file, by pushing the tang as far into the "grip" as it will go, and then turn the handle to the right until it is tight and firm.

SCREW FERRULE FILE HANDLE.

This file handle consists of a hard, smooth screw thread in the front end of the ferrule, and will cut its own way into the soft tang of the file, by pushing the handle as far on the tang as it will go and then screw it on until it is firm and tight. They are made in six sizes, but they will hold any size or shape that will go into the ferrule.

TESTIMONIALS.

Disston's Saws are used wherever lumber is made, and are the criterion of excellence by which the merits of all others are judged. In proof of this, we submit a few of the many testimonials which we receive daily from all parts of the world.

"CENTRE POINT, DOLBY CO., WEST VIRGINIA, *Aug. 12, 1886.*

"HENRY DISSTON & SONS.

"SIRS:—Your saw came all right. I have used it two weeks, and in that time sawed eighty thousand feet. It is the best saw I ever run—I would not give it for any solid-tooth saw made. It gives entire satisfaction, and I would advise all mill men to buy that kind, as it needs less gumming, and does not take half the work to keep it in order.

"Yours truly,

"J. N. DORSON."

"DUCK HILL, MISSISSIPPI, *Feb. 24, 1887.*

"MESSRS. HENRY DISSTON & SONS.

"GENTS:—I have been running a mill now 12 months under the instructions you sent me last March, and am prepared to say that it seems to me that if any thinking man will run according to your instructions, he can't help but succeed. I to-day retire from the mill business, and taking this method of thanking you for your advice, I will but say, in the future I shall use nothing but your best hand-saws in my trade. We have some excitement in iron ore here now and don't know what will be the result. Hoping and wishing you the very best success, I remain

"Yours obediently,

"H. N. KING."

"SOUTH WHITLEY, INDIANA, *Dec. 29th, 1886.*

"HENRY DISSTON & SONS.

"GENTS:—We have used a number of your saws, the last being a 70 inch. Each and every one has given entire satisfaction. We can cheerfully recommend them to all mill men.

"Yours truly,

"J. M. BRIANT & CO."

"PADUCAH, KENTUCKY, *Dec. 31st, 1886.*

"MESSRS. HENRY DISSTON & SONS.

"DEAR SIRS:—Yours of the 29th to hand, contents noted. The Gang Saws we purchased from you gave perfect satisfaction in every respect and we found them to be all you claimed for them.

"Yours truly,

'LANGSTAFF-ORM MFG. CO."

"CENTREVILLE, N. H., *Dec. 24, 1886.*

"HENRY DISSTON & SONS.

"SIRS:—The 52-inch Circular Saw I bought of you last December has been a good one, and I have taken one from your stock in Boston to run this winter.

"Yours truly,

"J. W. ROBERTS."

"YATESVILLE, N. C., *Dec. 28, 1886.*

"MESSRS. H. DISSTON & SONS.

"GENTLEMEN:—Yours of the 23d to hand, and noted. We have been using your chisel bit and solid tooth saws for the last five years, and find them satisfactory in every way.

"Yours truly,

"WHITEHOUSE & HAYES."

"BASS RIVER, N. S., *Dec. 30th, 1886.*

"MESSRS. HENRY DISSTON & SONS.

"DEAR SIRS:—We have had many years' experience with Disstons' saws, and find them more satisfactory than any others we have used.

"UNION FURNITURE AND MDSE CO., L'd.

"E. FULTON,

"Secretary-Treasurer."

"EASTMAN, GA., *1 Month, 3, 1887.*

"HENRY DISSTON & SONS.

"GENTS:—Yours, by Mr. Harper, at hand. I have been using Disstons' saws since 1859, and consider them the best solid saw in the market.

"Yours truly,

"J. T. COLCORD."

"WILLIAMSPORT, PA., *May 26, 1887.*

"MESSRS. HENRY DISSTON & SONS, Philadelphia, Pa.

"GENTLEMEN:—Your letter of the 24th asking our views as to the merits of your saw is at hand. We would say, we are using a 64-inch saw, 6 and 7 gauge with 72 teeth, running it with steam feed, and have run it without taking it off the mandrel three consecutive days, cutting from 50,000 to 55,000 feet per day. We cut 63,000 feet at one time without pointing. We have run it as high as 17-inch feed, without injuring the lumber.

"Yours truly,

"CORCORAN, RICHARDS & CO."

"CARLETON, ST. JOHN, N. B., *Dec. 29, 1886.*

"HENRY DISSTON & SONS.

"GENTLEMEN:—Yours of the 28th at hand. You can use our name as recommending the Disston saw as being of as good quality and more uniform than any make we have ever used.

"Yours truly,

"HAYFORD & STETSON."

"CHATTANOOGA, TENN., *January 1, 1887.*

"MESSRS. HENRY DISSTON & SONS.

"GENTS:—We have used your make of band saws for several years. We have had less trouble from breakage than formerly, and we cheerfully recommend them as superior to any we have ever used.

"Respectfully,

"JOHN HOWENSTINE & CO."

"*January 21st, 1885.*

"MR W. H. BLAKELEY, Melbourne.

"DEAR SIR ;—I have much pleasure in recommending Messrs. H. Disston & Sons' circular and vertical saws, and, during an experience of over thirty years, have found them superior to any others.

"I am, dear sir, yours truly,

"A. G. PARXICOTT, Manager for Beecham, Clark & Co.

"Phœnix Saw Mills."

"NORTH FITZROY, *Jan. 27th, 1885.*

"MR. BLAKELEY.

"I have much pleasure in stating I have used Messrs. Disston & Sons' circular and vertical saws for some time, and find them to give every satisfaction.

"Yours truly,

"CHARLES CRISPE."

"*January 27th, 1885.*

"TO MR. W. H. BLAKELEY, 116 Russell Street.

"Having been requested to give our opinion in reference to Disston & Sons' saws, we have much pleasure in stating that we consider them far superior in every respect to those of any other makers we have been in the habit of using. We prefer them to those of English manufacture, and would not use any others whenever a supply of them is available.

"HALSTEAD, KERR & CO.,

"Albion Saw Mills, Melbourne."

"BRUNSWICK, *Jan. 12th, 1885.*

"MR. W. H. BLAKELEY.

"DEAR SIR ;—We have much pleasure in testifying to the efficiency of Disstons' Saws. We have now used them, both circular and frame, for some considerable time, and find them superior to any we have before used, both in temper and regularity of gauge.

"We are, yours truly,

"COOK & SWINBOURN."

"*March 7th, 1885.*

"MR. W. H. BLAKELEY, Sawmaker,

"Russell Street, Melbourne.

"DEAR SIR :—The Disston & Sons' saws supplied by you, we have much pleasure in saying, are far and away the best circular saws we have ever used, and we have used many good makers' saws during the last five and twenty years in Adelaide.

"Yours truly,

"S. HARVEY & WM. KING.

"Adelaide."

"*February 27th, 1885.*

"TO MR. W. H. BLAKELEY, Russell Street, Melbourne.

"DEAR SIR ;—Having been requested to give our opinion respecting your saws, we do so with great pleasure, having had them in use for six years, and have been thoroughly satisfied. We recommend them to all we know.

"We remain, yours, &c.,

"EASON & CLAYTON.

"Brighton & St. Kilda Saw Mills."

"EMPORIUM, PA., *Dec. 27th, 1886.*

"HENRY DISSTON & SONS.

"GENTS;—Your favor of the 23d inst. at hand; in reply would say will be pleased to help you in this matter, as I can speak from a 20 years' experience of handling your saws, and have found them to give good satisfaction in every respect, especially your gang saws. Please forward me one of your new circulars or catalogues when ready for issue.

"Yours truly,

"HENRY AUCHER."

"COLD BROOK SPRINGS, MASSACHUSETTS, *Dec 15th, 1885.*

"MESSRS. HENRY DISSTON & SONS.

"SIRS:—The 48-inch chisel point saw, 10x11 gauge, with 36 teeth, that I bought of you March 5th, 1884, is the best saw that I ever run for a thin saw. I cut out 300,000 oak and pine, and 200,000 railroad sleepers with one set of teeth, and did not spend over ten hours time in filing, while sawing that quantity of lumber. I think that is doing well with one set of teeth. I see in the *Lumber World* that you have got up a patent tooth, and if it is better than your chisel-tooth, please let me know.

"Truly yours,

"R. F. PARKE."

"WESTONIA, GEORGIA, *Jan. 5th, 1887.*

"MESSRS. HENRY DISSTON & SONS.

"GENTS:—All we say about the Disston Saw is that we prefer them to any saw that we know of. At times we cut a good deal of turpentine timber that has turned into hard light-wood, and we find that they stand up to this work better than any saw we have ever used.

"Respectfully,

"S. R. & J. D. WESTON."

"EDWARDS, NEW YORK, *Dec. 20th, 1885.*

"HENRY DISSTON & SONS.

"GENTS;—Please send me your Illustrated Catatogue and price list of your circular saws,

"And oblige, "GEO. W. FLACK.

"P. S.—The chisel-tooth saw I got last Spring (through the Birdsall Co.,) works splendid."

"CLEVELAND, OHIO, *Jan. 20, 1883.*

"HENRY DISSTON & SONS.

"GENTLEMEN:—Having used the 60 and 72-inch diameter, Disston Patent Chisel Point Circular Saws, purchased of you, for seven months, we have no hesitation in saying that they are the best we have ever used. The chisel points furnished with same are a perfect success, and often enable us to cut 300,000 feet with one set of points. We expect to use one more circular saw this season, and possibly two, and as soon as we can arrive at a decision will place the order with you for the Disston.

"Respectfully yours,

"CLEVELAND SAW MILL & LUMBER CO.

"George R. Payne, Vice-President."

"ORANGEBURG, SOUTH CAROLINA, *July 28, 1886.*

"MESSRS. HENRY DISSTON & SONS.

"SIRS:—The chisel-point saw arrived on Monday, the 23rd, and was put to work immediately. It works admirably, walking through 36-inch logs, cypress and pine, without trouble, and making as beautiful lumber as I have ever seen.

"Yours truly,

"F. COPES."

"MINNEAPOLIS, MINNESOTA, *Dec. 29th, 1886.*

"MESSRS. HENRY DISSTON & SONS.

"SIRS:—I have used the Disston saw in my saw-mill for the last seven years, and they have given me better satisfaction than any other make that I have used, and I have tried almost all brands of saws.

"Yours truly,

"JAMES GOODNOW."

"HENDERSON, KENTUCKY, *Dec. 25th, 1886.*

"HENRY DISSTON & SONS.

"SIRS:—The 62-inch 10-gauge saw you made for P. J. McNamary, eighteen months ago, ordered by me, is running yet, as good as the first day I hung it on the mandrel. I have been running it myself ever since I got it, and have never had a hammer on it. I believe I can wear it out and never have it repaired. I run it at the 3-inch feed, and it never has been warm one particle since I got it. It is the finest saw that I ever put a file on. I will want another one after a while. You make the boss saw to my notion, and the man that can't run one of them is no sawyer.

"Yours truly,

"LEON WORLEY, Sawyer"

"RONCEVERTE, WEST VIRGINIA, *Jan. 4th, 1887.*

"MESSRS. HENRY DISSTON & SONS.

"GENTLEMEN:—We have used your circular and gang saws for years with satisfaction, and take pleasure in recommending them to the public as first class in every respect.

"Yours very truly,

"E. C. BEST,

"General Manager."

"LAGRANGE, OHIO, *Jan. 29th, 1883.*

"HENRY DISSTON & SONS.

"GENTLEMEN:—In regard to the 60-inch chisel tooth saw, 8 gauge, of H. Disston & Sons' make, which we bought of you, we were well pleased with it. It stands up good under 3-inch feed with full cut of saw (27-inch cut) in frozen timber. We are cutting from 8,000 to 12,000 feet of lumber per ten hours, edging all lumber with same saw. We could nearly double the amount if we had a good edger. We saw all kinds of timber common to Northern Ohio, such as oak, ash, beech, maple, hickory, elm, poplar, basswood, etc.

"Three-inch being the largest feed we have, we do not know how much our saw would stand, but would no doubt carry much heavier feed, at least in soft wood. We are cutting from 100,000 to 200,000 feet with each set of 36 teeth. We have used one set of teeth three weeks after the corners were worn off, by spreading with swage, which works well if the swage is of suitable shape. You may refer to us any one you wish in regard to Disston's saws.

"Yours truly,

"BUTLER & BEHNER."

"CONROE, TEXAS, *Dec. 31st, 1886.*

"HENRY DISSTON & SONS.

"SIR:—In answer to yours of the 29th to give testimonial as to the H. Disston's saws, I will state that I have used the Disston saws for the last twelve years, and that I find them the best that I have ever used. I have tried a number of other makes, but I think the Disston far superior to any that I have ever tried.

"Respectfully yours,

"I. CONROE."

"ORLANDO, FLORIDA, *Aug. 26th, 1885.*

"Send two dozen eyes or holders for teeth for your inserted tooth saw, 7 gauge, 52-inch, 32 teeth. Your saw works very fine in our hard yellow pine. Prefer it to any we have used.

"ORLANDO LUMBER CO.,
"W. R. GULICK."

"HOUSTON, TEXAS, *Dec. 30th, 1886.*

"MESSRS. HENRY DISSTON & SONS.

"SIR:—I cheerfully say that the Henry Disston & Sons' circular saws I purchased of you are the best I ever used, and in future no other shall be my choice.

"Yours truly,
"HERMAN A. BAUER."

"CYNTHIANA, KENTUCKY, *Feb. 16th, 1885.*

"MESSRS. HENRY DISSTON & SONS.

"GENTS;—I bought of you through W. Baldridge & Co., of Cincinnati, Ohio, about one year ago, one of your 52-inch, 6x7 gauge, patent chisel point circular saws. I have run it ever since through the roughest burr oaks and sugar tree, and in fact as rough timber and as tough as the Blue Grass Region of Kentucky affords, and it cost me $85.50, and to-day I would not take $500.00 for it, if I could not replace it. It works like a charm, no trouble to keep in order, always exactly round and each tooth cuts its proportion; it has not cost me one cent yet, and I can make smoother lumber with less pains than with any other make and can saw off more points of dogs, and nails, and iron of any description, than any other make, and do less injury. I actually sawed half off the lower half of my saw guide a few days ago, and did not hurt the saw a particle. Broke one tooth, and I just put in a new set in twenty minutes and went on, whereas if it had been a solid tooth it would have cost me probably ten dollars. They will stand more hardships than any other saw made in U. S. A.

"Yours truly,
"J. E. HAYES."

"BARNUM, TEXAS, *Jan. 1st, 1887.*

"HENRY DISSTON & SONS.

"SIR:—The Disston saws bought of you some time ago have given perfect satisfaction, and we have almost decided not to use any other. We have not had occasion to use any of Para belting yet, will write you when we have thoroughly tested it.

"Yours truly,
"W. T. CARTER & BRO."

"SAWMILLS, NARRACOORTE, S. A., *Jan. 15th, 1885.*

"MR. W. H. BLAKELEY, Sawmaker, Russell Street, Melbourne.

"DEAR SIR:—I have for the past five years used H. Disston & Sons' circular and vertical saws, and during that time I have found them to give complete satisfaction, having had them at the roughest of work both in South Australia, New South Wales, and Victoria. I am taking care to recommend them to my friends and sawmill men generally with whom I am acquainted; and as I find them of such a superior temper and so much more reliable than any other maker I have tried, I will not use any other when a "Disston" is procurable.

"I am, dear sir, yours faithfully,
"A. C. MUNRO."

"SURRENCY, Georgia, *Jan. 7, '86.*

"HENRY DISSTON & SONS,

"GENTS:—The Disston saws purchased from you some time ago are doing all we would ask, and in fact, my saw filer prefers them to any saws we have now in use. In future, all saws I order will be the Disston saws, as I believe them to be the cheapest and best saws that saw-mill men can use.

"Yours truly,

"GEO. L. IFFERMAN,

"Sup't McDonough & Co."

"WILLIAMSPORT, Pa., *June 1, 1887.*

"MESSRS. HENRY DISSTON & SONS, Philadelphia, Pa.

"GENTLEMEN:—It affords us much pleasure to inform you that the two 66-inch, 8 gauge saws, with eighty-five teeth, are the best saws we ever used. We have the —— solid tooth saw, same size and gauge, also —— inserted tooth saw, but we do not use them only as a change, as we can make better lumber, and a great deal more of it with your saws than with any of the others. We run from 6- to 20-inch feed with your saw. Have used your gang saws for over fifteen years, and consider them the best in the market.

"Yours truly,

"BROWN, CLARK & HOWE."

"DOUGLASTOWN, N. B., *Dec. 26, 1886.*

"MESSRS. HENRY DISSTON & SONS.

"GENTLEMEN;—Yours of yesterday received. I do not now use any other saw than Disston's in my gangs, and *find them good*—in fact, they are the only saw in my experience which will stand the work. For gang saws give me Disston's every time. It does not pay to touch any other.

"Yours truly,

"E. HUTCHISON."

"SUFFOLK, VIRGINIA, *Dec. 29, 1886.*

"MESSRS. HENRY DISSTON & SONS, Philadelphia, Pa.

"DEAR SIRS:—The three 64-inch circular saws bought of you a short time since for our new mill, made of your compressed steel, have given entire satisfaction.

"They swage better, and will do more work from one filing than any saws we have ever seen.

"Yours truly,

"THE GAY MANUFACTURING CO."

"REHN, PA., *June 1, 1887.*

"MESSRS. HENRY DISSTON & SONS.

DEAR SIRS:—I don't know as I ever told you how I liked your Band Saws; but will say now they are the best there are, without a doubt.

"Yours truly,

"F. J. McNUTT."

Part II.

HENRY DISSTON & SONS'

[INCORPORATED.]

KEYSTONE

Saw, Tool, Steel & File Works,

BRANCH HOUSE,
Chicago, Ill.

TRADE MARK

BRANCH HOUSE,
Louisville, Ky.

PHILADELPHIA, PA.

PART II.

Construction of Saws

And How to Keep Them in Order.

INTRODUCTION.

The demand for a book of instruction on saw filing having demonstrated itself to us not only by personal inquiry and letter, but also by the return of fine quality saws, pronounced defective through a lack of knowledge of how to keep them in order, or by the use of extensively advertised so-called saw-sets and other tools,—which pull the saw blade apart or so distort it as to render it unfit for use—has led us to compile this little work for gratuitous distribution for the enlightenment of the amateur and the improvement of the expert mechanic.

Having thus introduced the object of this book, we will endeavor to give in the following pages such practical information as to the proper methods of keeping saws in order and of the tools with which to do so, that will overcome the above mentioned pit-falls to the proper working of the saw. We offer our large experience and the reputation of our goods for the efficiency of this treatise, which has been gleaned from the most practical and oldest saw filers of our establishment, many of whom have been with us two score of years. While we admit there are other methods of putting saws in order, there is no difference in the result obtained; our modes we claim the easiest and equally or more effective.

We take occasion to thank our patrons for their appreciation of the high standard that our products have attained,—which has been the basis for our constantly increasing business—which shall be maintained, and trust thereby to renew our good will and increase our trade in future.

HENRY DISSTON & SONS, Incorporated.

CONTENTS.

PART II.

PRINCIPLE OF CONSTRUCTION.

The saw is either reciprocating or continuous in action, the first being a flat blade and practically straight edge, making a plane cut, as in hand, mill, jig and sash saws; the latter, either a circular or rotating disc, cutting in a plane at a right angle to its axis, a cylindrical or barrel shape with a convex edge cutting parallel to its axis, or a continuous ribbon or band running on two pulleys making a plane or curved cut with a straight edge parallel to their axes of rotation. Practically speaking, the teeth are a series of knives set on a circular or straight line, each tooth cutting out its proportion of wood and kept from cutting more by the teeth on either side of it. Each tooth should cut the same amount and carry out the chip or dust, dropping it to the sides or below the material being sawed. Different kinds of wood require teeth different in number, angle or pitch and style of filing.

The perfection of a saw is one that cuts the fastest and smoothest with the least expenditure of power; to do this, it is evident that each tooth should be so constructed and dressed as to do an equal proportion of the work, for if any of the teeth are out of line or shape, they are not only useless themselves, but a disadvantage to the others. We find many good mechanics who frankly acknowledge that they never could file a saw satisfactorily; the probable reason is that they never studied the principle of the action or working of the tool. There is no reason why any man of ordinary mechanical ability should not be able to put and keep his saw in order, but like all trades, it requires practice and study of the subject.

A careful study of the following illustrations and explanations, will greatly assist in the selection of a saw and the best methods of keeping it in proper working order.

A saw tooth has two functions—paring and scraping. A slitting or ripping saw for wood should have its cutting edge at about right angles to the fibre of the wood, severing it in *one* place, the throat of tooth wedging out the piece.

In a cross-cut wood saw, the cutting edge also strikes the fibre at right angles to its length, but severs it on *each side* from the main body before dislodging it.

RIP SAWS.

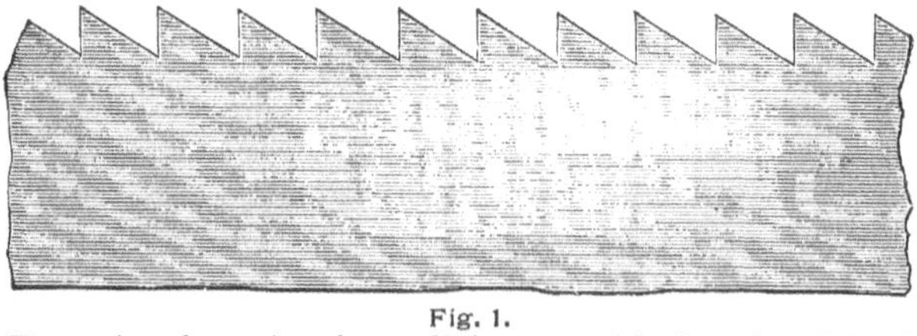

Fig. 1.

Fig. 1 is a four-point rip or slitting saw with the rake all in front, where the cutting duty is. This saw should be filed square across, filing one-half the teeth from each side after setting, which will give a slight bevel to the cutting edge of tooth, as it should be for soft wood; for medium hard woods a finer toothed saw with five points to the inch should be used and dressed in the same manner; for the very hardest and toughest cross-grained woods a saw still finer, the teeth filed slightly beveling, as ripping cross-grained stuff

Fig. 2.

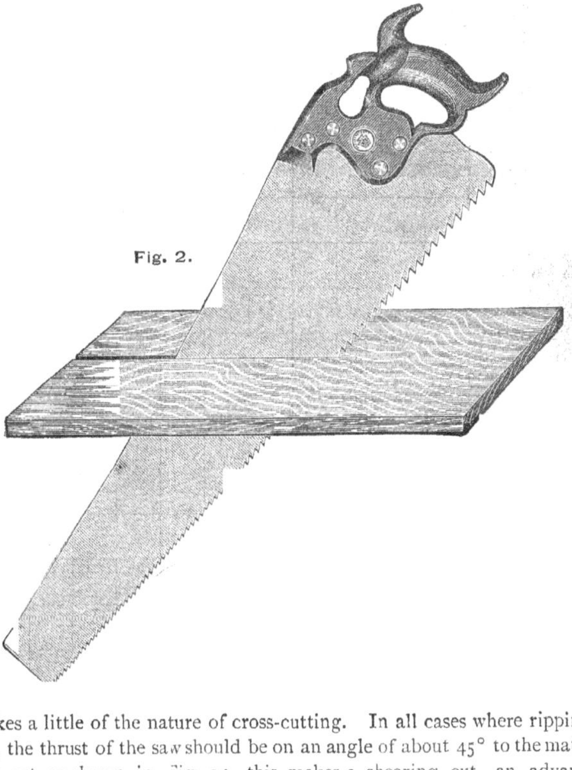

partakes a little of the nature of cross-cutting. In all cases where ripping is done, the thrust of the saw should be on an angle of about 45° to the material being cut, as shown in Fig. 2: this makes a shearing cut, an advantage that can be very quickly demonstrated with an ordinary pocket knife cutting any piece of wood. For ripping thoroughly dry lumber, it will be found advantageous to use an extra thin back saw which will run without set.

CROSS-CUT HAND SAWS.

In cross-cutting, the fibre of the wood is severed *twice*—on each side of the saw—the thrust dislodging and carrying the dust out.

Fig. 3 is a five-point peg tooth cross-cut saw with the rake on the side. For the same reason that the rip saw has the rake on front of tooth, the cross-cut has it on the side, as that is where the cutting duty is. The bevel or fleam to teeth in Fig. 3 is about 45°, while there is no pitch at all;

Fig. 3.

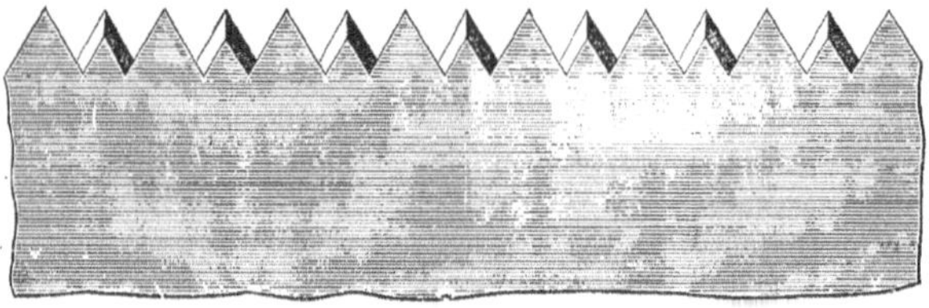

the angle on each side being the same, forms the "*peg tooth*," which is best adapted to cutting soft, wet and fibrous woods, and used principally as a buck saw.

In all cases, the size and length of teeth depend largely upon the duty required; a long tooth has the demerit of being weak and liable to spring, but the merit of giving a greater clearance to the saw-dust. The throat space in front of each tooth must be large enough to contain the dust of that tooth from one stroke; the greater the feed, the deeper the dust chamber required, or, more teeth. Where the teeth are fine, the shape of the throat is of special interest.

The first point to be observed in the selection of a saw is to see that it "hangs" right. Grasp it by the handle and hold it in position for working, to see if the handle fits the hand properly. These are points of great importance for comfort and utility. A handle should be symmetrical, and the lines as perfect as any drawing. Many handles are made of green wood; they soon shrink and become loose, the screws standing above the wood. We season our handle-wood three years before using. An unseasoned handle is liable to warp and throw the saw out of shape. The next thing in order is to try the blade by springing it, seeing that it bends regularly and evenly from point to butt in proportion as the width and gauge of the saw varies. If the blade is too heavy in comparison to the teeth, the saw will never give satisfaction, because it will require more labor to use it; the thinner you can get a stiff saw the better; it makes less kerf and takes less muscle to drive it. This principle applies to the well-ground saw. There is less friction on a narrow true saw than on a wide one; you will get a smaller portion of blade, but you will save much unnecessary labor at a very little loss of the width.

See that it is well set and sharpened and has a good crowning breast; place it at a distance from you and get a proper light on it, by which you can see if there is any imperfection in grinding or hammering. We should invariably make a cut before purchasing a saw, even if we had to carry a board to the hardware store. We set our saws on a stake or small anvil with a hammer; a highly tempered saw takes three or four blows, as it is apt to break by attempting to set it with but one blow. This is a severe test, and no tooth ought to break afterwards in setting, nor will it, if the mechanic adopts the proper method. The saw that is easily filed and set is easily made dull. We have frequent complaints about hard saws, but they are not as hard as we would make them if we dared; but we shall never be able to introduce a harder saw until the mechanic is educated to a more correct method of setting it. As a rule, saws are set more than is necessary, and if more attention was paid to keeping points of teeth well sharpened, any well-made saw would run with very little set, and there would be fewer broken ones. The principal trouble is that too many try to get part of the set out of the body of the plate, while the whole of the set should be on the teeth. Setting below the root of the tooth distorts and strains the saw-plate, which may cause a full-tempered cast-steel blade to crack and eventually break at this spot, and is always an injury to the saw, even if it does not crack or break.

The teeth of a hand-saw should be filed so true that, on holding it up to the eye and looking along its edge, it will show a central groove down which a fine needle will slide freely the entire length; this groove must be angular in shape and equal on each side, or the saw is not filed properly and will not run true.

Fig. 4.

Fig. 5.

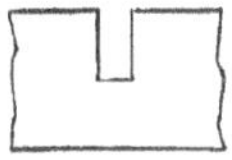

Fig. 6.

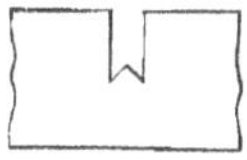

Fig. 7.

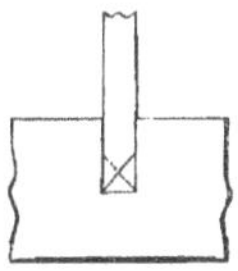

Fig. 4 shows how the groove should appear on looking down the edge of the saw; the action should be such that the bottom of kerf will present the appearance as shown in Fig. 5, and not like Fig. 6; the cutting action is as shown in Fig. 7, the cutting being done with the outside of tooth, the fibre of the wood is severed in the two places and the wood is crumbled out from point to point by the thrust of saw.

The proper amount of bevel to give the teeth is very important, as is demonstrated by the above figures, for if too much bevel is given, the points will score so deeply that the fibres severed from the main body will not crumble out as severed, but be removed by continued rasping, particularly in hard woods, as they require less bevel, as well as pitch, than soft wood.

Fig. 8 on next page, shows a six-point cross-cut saw filed with a medium amount of bevel on front or face of tooth, and none on the back. This tooth is used in buck saws, on hard wood, and for general sawing of woods of varying

Fig. 8.

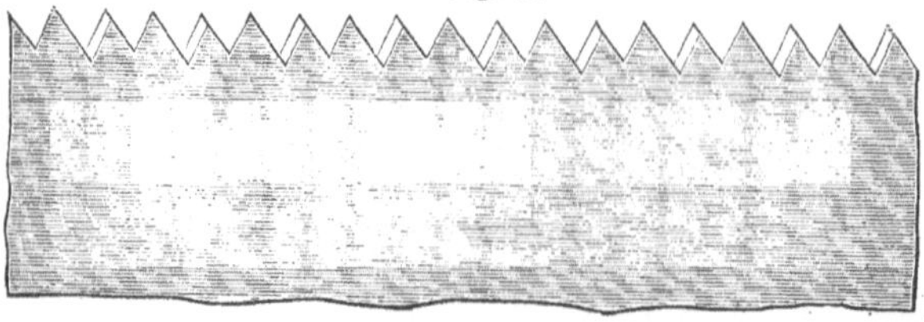

degrees of tenacity. This style of dressing is the best, but a number of saws each having teeth suited to its particular work, will be found more advantageous than trying to make one saw serve for all kinds of hand saw work.

We will now consider the cross-cut saw tooth, in regard to rake or pitch; this being one of the most important features, too much care cannot be taken to have the correct amount of pitch for the duty required. To illustrate this, Fig. 9 represents a board, across which we wish to make a deep mark or score with the point of a knife; suppose we hold the knife

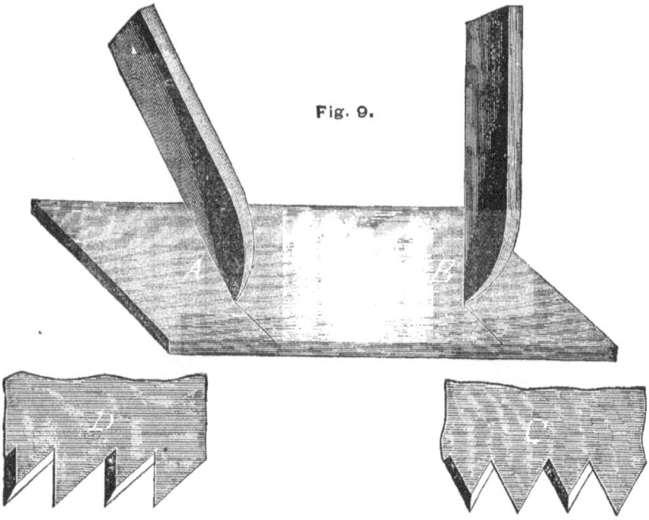

Fig. 9.

nearly perpendicular as at *B*, it is evident it will push harder and will not cut as smoothly as if it was inclined forward as at *A*; it follows then that the cutting edge of a cross-cut saw should incline forward as at *C*, rather than stand perpendicular as at *D*.

Too much hook or pitch, and too heavy a set are very common faults, not only detrimental to good work but ruinous to the saw; the first, by having a large amount of pitch, the saw takes hold so keenly that frequently it "*hangs up*" suddenly in the thrust—the result, a kinked or broken blade; the second, by having too much set, the strain caused by the additional and unnecessary amount of set is out of proportion to the strength of the blade, and is broken in the same manner. The most general amount of pitch used is 60°, though this may be varied a little more or less to advantage, as occasion may demand.

The next point to be considered is the bevel, or fleam, of the point. In Figs. 10, 11 and 12, the filer, as in all cases, files from the heel to the point, which is the only correct way. The file is supposed to be horizontal to

Fig. 10.

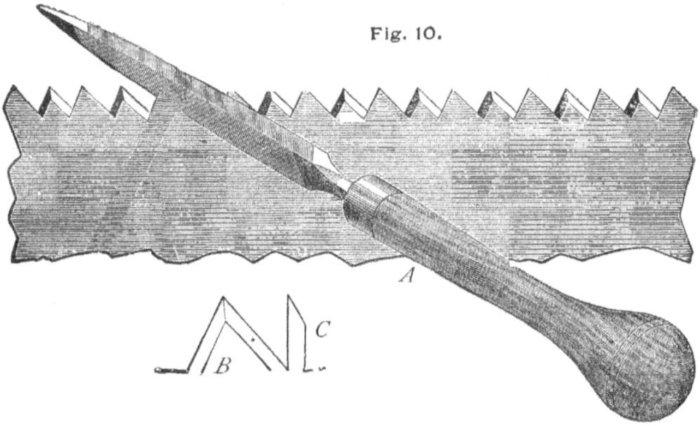

Fig. 11.

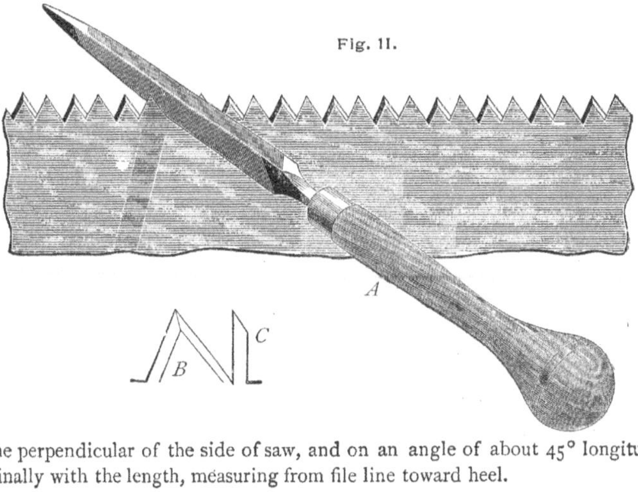

the perpendicular of the side of saw, and on an angle of about 45° longitudinally with the length, measuring from file line toward heel.

Fig. 10 is a five-and-a-half-point cross-cut saw showing the same amount of fleam front and back; this saw is best suited for work in soft wood, and where rapid, rather than fine work is required. *A* shows the position of the file, *B* an exaggerated view of shape of point, and *C* the shape of point.

Fig. 11 is a seven-point saw for medium hard woods, illustrated in same manner as Fig. 10. This tooth has less fleam on the back, which gives a shorter bevel to point, as at *C*.

Fig. 12.

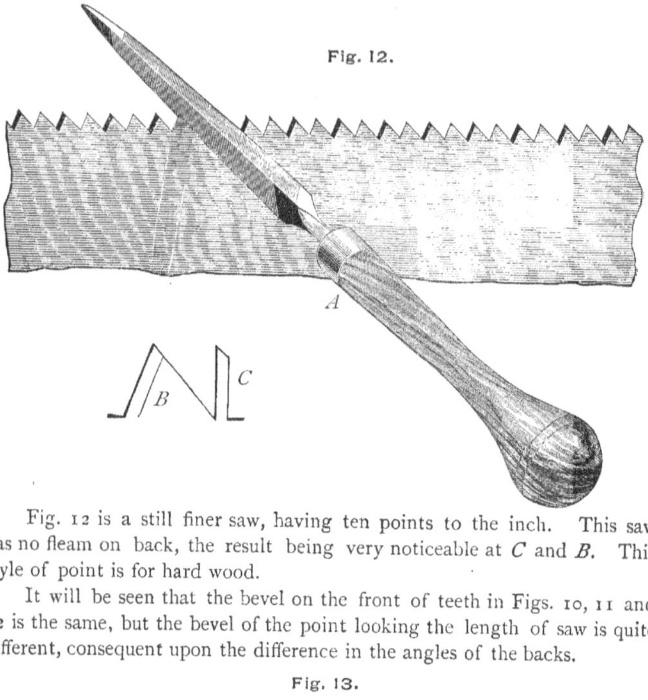

Fig. 12 is a still finer saw, having ten points to the inch. This saw has no fleam on back, the result being very noticeable at *C* and *B*. This style of point is for hard wood.

It will be seen that the bevel on the front of teeth in Figs. 10, 11 and 12 is the same, but the bevel of the point looking the length of saw is quite different, consequent upon the difference in the angles of the backs.

Fig. 13.

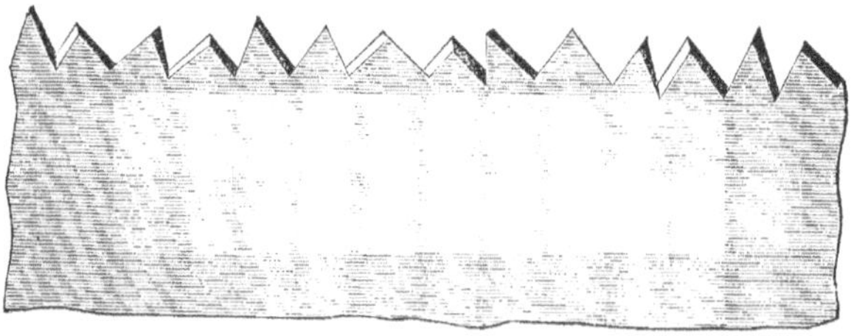

Fig. 13 is a representation of some of the saws we have seen; there are entirely too many such now in use, and we have no doubt their owners are shortening their lives in the use of them as well as those of the saws. To owners of such saws we say, take them to the factory and have them re-toothed, or buy a new saw and take a fresh start, and steer clear of this style of filing.

As we said in the preceding pages, and as will be seen by Figs. 10, 11 and 12, the filing should be done from the heel of saw toward the point. Many practical saw filers contend this is wrong, that the filing should be done from point of saw toward the handle, but the only support they have for their theory is that they do away with the feather edge that the filing from the heel of saw puts on the cutting face of tooth. The feather edge is no objection, as the main part of it is removed when the teeth are side-dressed after filing, as we direct in our summary of saw filing on page 16. Against the correctness of filing from point to handle may be cited the following objections:

Where a different angle of back is required, (it being remembered that angle of face should be the same in nearly all cross-cut hand saws, and that angle of back governs angle of point,) it will be found very difficult to obtain it without changing angle of face of tooth, and as the cutting duty is on the long side of face, any change is of course of great influence.

Again, (though we think the above argument sufficient) to file from point of saw, it is necessary to file with the teeth bent toward the operator; this will cause the saw to vibrate or chatter, which not only renders good, clean, even filing impossible, but breaks the teeth off the file.

In the preceding illustrations, we have only given the coarser saws that are in most general use, but the same principle of filing should be applied to the finer toothed saws regarding angles and pitch suitable for woods of different degrees of hardness, the only actual difference being that one saw has finer points, and they being finer, require a little more care and delicate touch in setting and filing.

Fig. 14.

Fig. 14 is a section of an eleven-point saw suitable for the finer kinds of work on dry, soft woods, such as cutting mitres, dove-tailing, pattern work, etc.

Fig. 15 shows a section of saw with same number of points as Fig. 14,

Keystone Saw, Tool, Steel, and File Works,

PHILADELPHIA, PA.

To Users of Circular, Mulay, Mill and Gang Saws:

We desire to call special attention to the superior quality of our GANG, MILL AND CIRCULAR SAWS. We have made recent valuable and important improvements in the manufacture of these saws. The steel is made expressly for the purpose. The tempering is done by an improved process which insures that the swaging can be done without the risk of the teeth splitting, crumbling, or spalling off. These saws are ground by a new method, giving perfect regularity of gauge, thus enabling parties to turn out lumber of uniform thickness, and the saws will run more freely and with less power than any others made; they are also hammered by a patent process of our own, making them uniformly of the same tension throughout the body of the saw from the edges into the centre in such a manner as to allow a greater amount of strain and wear without becoming long or stretched on the tooth edge. Attached hereto will be found blanks for ordering. If full and proper instructions are given, we guarantee satisfaction. Should a saw prove defective in any of the above particulars, a new one will be given in its place if returned in sixty days.

Yours respectfully,

HENRY DISSTON & SONS, Incorporated.

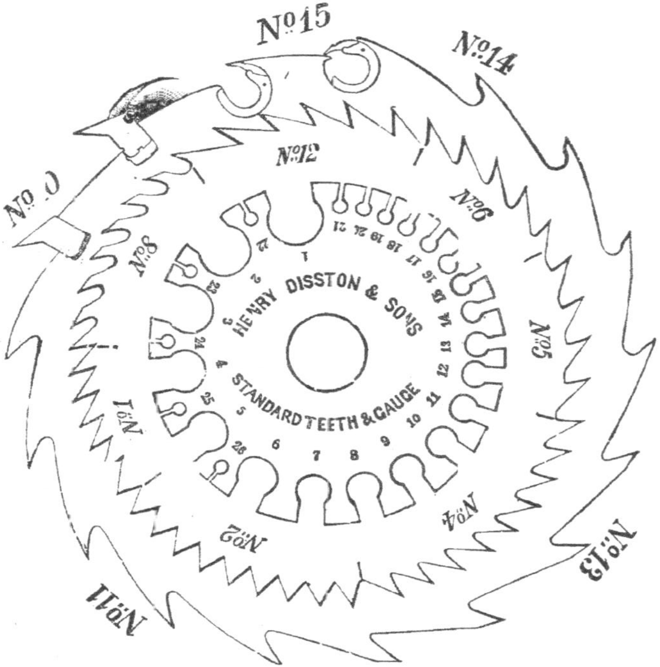

BLANK FOR ORDERING CIRCULAR SAWS

RIGHT HAND SAW. FROM LEFT HAND SAW.

HENRY DISSTON & SONS

Philadelphia, Pa.

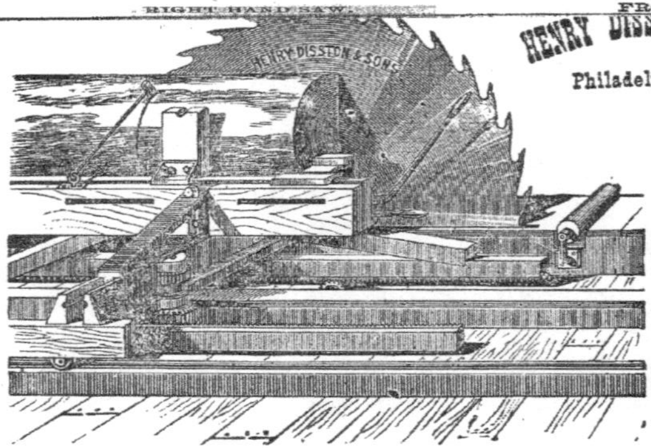

Diameter of Saw, ... *inches.*	*Pin Holes,* .... *inches.*
Thickness of Saw at rim, ... *gauge.*	*Distance between pin holes from centre to centre,* .... *inches.*
Thickness of Saw at centre, ... *gauge.*	*Right or left hand Saw, see engraving above,* ....
Number of Teeth in Saw, ...	*Greatest feed at each revolution of Saw,* .... *inches.*
Kind of Tooth and No.	*Kind of Lumber to be sawed,* ....
Whether spring or spread set, ...	*Speed of Saw, about* .... *revolutions per minute.*
Mandrel Hole, ... *inches.*	

All Saws With our Brand FULLY WARRANTED.

BRANCH HOUSES,
CHICAGO, ILL.
LOUISVILLE, KY.

HENRY DISSTON & SONS,
Incorporated
PHILADELPHIA, PA.

DIRECTIONS FOR SHIPPING

By .... *Express.*
By .... *Steamship.*
By .... *Regular Freight.*

Signed, ____________

From HENRY DISSTON & SONS, Incorporated, Philadelphia, Pa.

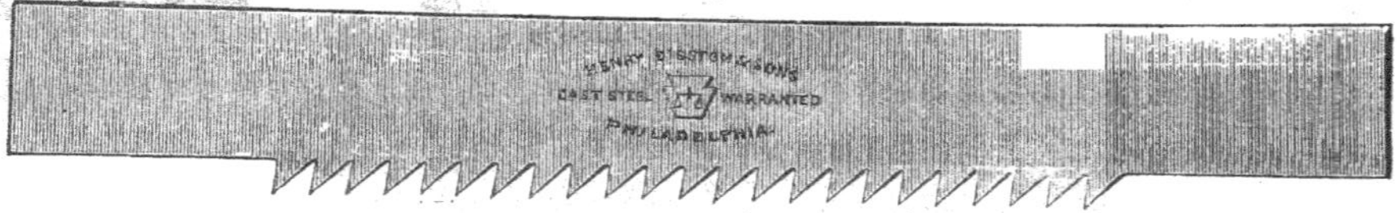

Length, ...	*Distance from bottom of Saw to point of front tooth,* ....
Width, ...	*Width at ends,* ....
Gauge, ...	*If tabbed, give exact distance between tabs, also thickness of Saw and tabs when finished,* ....
Space of teeth, ...	
No. of teeth, ...	
Hook of teeth, ...	*Kind of tabs wanted,* ....

All Saws with our Brand FULLY WARRANTED.

BRANCH HOUSES.

Randolph and Market Streets,
CHICAGO, ILLS.

923 West Main Street,
LOUISVILLE, KY.

HENRY DISSTON & SONS,
Incorporated
PHILADELPHIA, PA.

DIRECTIONS FOR SHIPPING.

By .... *Express.*

By .... *Steamer.*

By .... *Regular Freight.*

Signed, ...

Fig. 15.

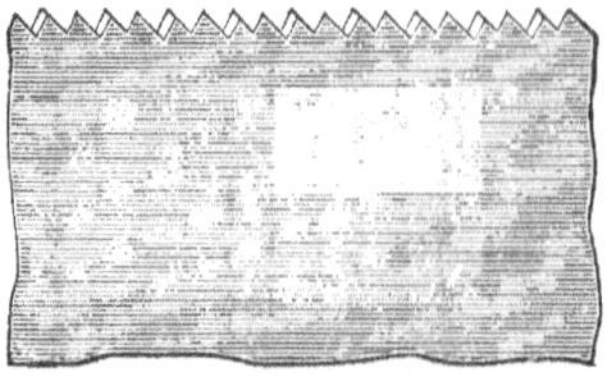

but filed same as Fig. 12. This saw is for finer work, same as Fig. 14, only on the medium hard woods.

Fig. 16.

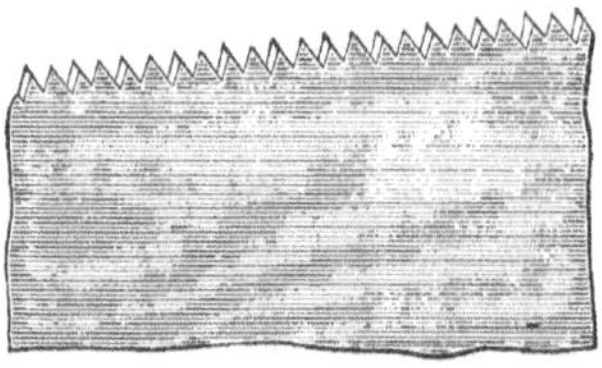

Fig. 16 is a still finer saw for fine work on the very hardest woods, having same dress as Fig. 14.

Fig. 17.

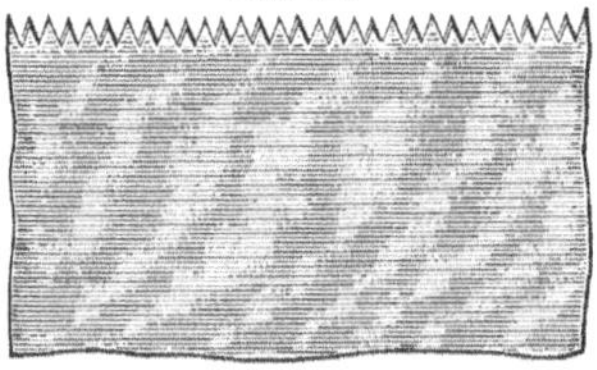

Fig. 17 is the finest toothed saw that is made for wood. All the above mentioned saws in Figs. 14, 15, 16 and 17, are made especially hard and will not admit of setting, but being made thinner at the back, when properly filed, will cut clean and sweet. Teeth such as shown in Fig. 17 are used principally on back saws, and are filed same as any hand-saw.

Fig. 18.

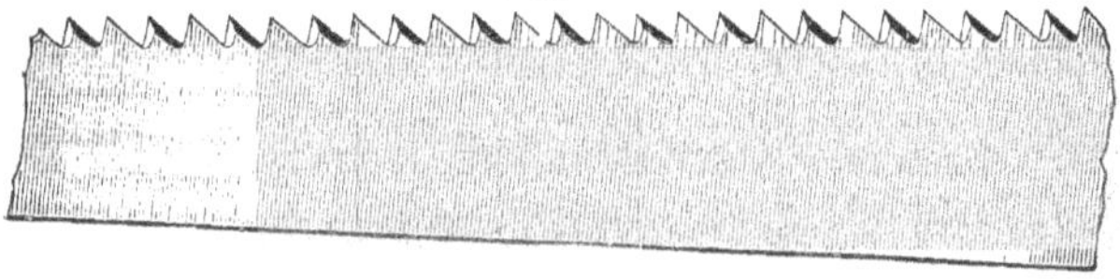

Fig. 18 is a section of a pruning saw which differs from a cross-cut hand-saw in being thicker, having a little more pitch to the teeth and being ground thinner on the back in proportion to its width. These, of course, are made for cross-cutting only, as there is not as great a variety in the work, nor as much difference in the woods to be awed as to degrees of hardness, being used only as a pruning saw on fruit nd shade trees, which are always practically green and comparatively soft. The illustration on page 13 shows number of points, pitch and bevel most generally used and best adapted to such work.

COMPASS SAWS.

These saws are for miscellaneous sawing, having in turn, cross-cut, rip and mitre. The best form of tooth for this purpose is the same as Fig. 18, excepting that it has a trifle less bevel. As the nature of the work partakes about as much of cross cutting as of ripping, and as a cross-cut saw will rip better than a rip will cross-cut, it is apparent the shape of tooth should be between the two. These saws are all ground thinner at back but set same as any hand-saw.

Scroll and web saws are ground, filed and set in the same manner, and should have pitch according to the work to be done. If more ripping than cross-cutting is done, as in large felloes, more pitch is given than in compass saws and *vice versa*, though these saws are almost universally run with a rip-saw tooth and have very little variation in the pitch.

BUTCHER SAWS.

These saws are for cutting bones. The pitch and number of points are about the same as a fine tooth hand-saw for medium hard wood, but filed straight through without fleam or bevel to teeth, with light, even set, same as in fine hand-saws.

HACK SAWS.

These saws are for cutting metal, such as brass, iron, or untempered steel, and should have a little finer tooth than the average butcher saw, being made to run without set and ground thin on the back for clearance. They are so hard that none but the best superfine files will sharpen them. Like the butcher saws, the filing must be straight through and no bevel.

SETTING SAWS.

This is an important part of the work of keeping a saw in order and should always be done *after* the teeth are *jointed* and before filing. In all cases the set should be perfectly uniform, as the good working of the saw depends as much on this as on the filing. Whether the saw is fine or coarse, the depth of set should not go, at the most, lower than half the length of the tooth, as it is certain to spring the body of saw if not break the tooth out. Soft, wet woods require more set as well as coarser teeth than dry, hard woods. For fine work on dry woods, either hard or soft, it is best to have a saw that is ground so thin on the back that it requires no set; such saws are made hard and will not stand setting, and an attempt to do so would surely break the teeth out.

There are many saw-sets that ruin the saw; the best form is one that involves the principle of the hammer and anvil; with such a set the teeth would all be bent evenly, and cannot be otherwise, though repeated blows be given. In the Star saw-set, represented in the following engraving, this principle is involved, and we guarantee this tool to do the work satisfactorily.

THE STAR SAW-SET.

Prominent among the advantages claimed for this set is that it can be operated by the foot by means of a treadle, thus leaving the hands free to guide the saw; or it can be used by striking on the top with a light mallet.

A is the plunger, operated by a treadle attached to *E*, under the machine, a slight tap with the foot setting the tooth; *B*, the hammer or striking part; *C*, the anvil; *D*, the movable gauge; *F*, the screw to regulate the amount of set. The striking part, and the anvil, or portion which receives the blow, are star-shaped, and similar in construction. The points are all of different sizes, numbered from one to six, and are designed to set different sized teeth. It will strike a blow as sharp and effective as though by a hammer, and is the most useful and complete saw-set that has ever been offered. If the saw is hard, several blows should be given in setting it, raising the back of the saw from the guide-screw *F* when the first blow is given, and gradually lowering it with each blow until the process is complete; thus many a good saw will be saved from utter ruin. A trial will suffice. Be sure to clean the saw teeth before setting.

ADJUSTABLE SAW CLAMP.

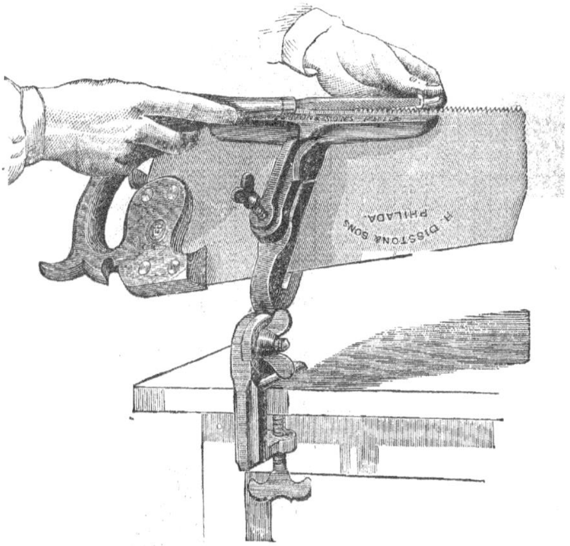

Care should be taken in filing a saw to keep the teeth of uniform size—not one large and one small, one up and one down. Unless your teeth are regular, your set can never be regular. When the teeth of a saw become irregular in size, it is useless to attempt to regulate them without filing them down until all the teeth are of equal height. Then proceed to regulate the size by filing straight through. We know from experience that not one man in a thousand, be he ever so practical and proficient, can regulate the teeth of a saw without first filing down and then filing straight through. After filing the saw, properly set and sharpened, lay it flat on a true board, rub over the points of the teeth on the side with a smooth or partly worn flat file, which will regulate the set and insure smooth cutting, making the filing last longer.

After this operation, should the saw not run true, take another cut with the file over the side toward which it leads. A fast cutting cross-cut saw should have deep teeth. To make them deep they must be filed on an angle; to do this to advantage the clamp (see engraving) should be used, and thus a deep gullet tooth can be filed as readily as a square bottom tooth. Each kind of work requires a tooth to suit it if good and quick work is wanted; hard wood requires a tooth with less rake than soft wood. Teeth generally have not enough saw-dust chamber.

LARGE CROSS-CUT SAWS.

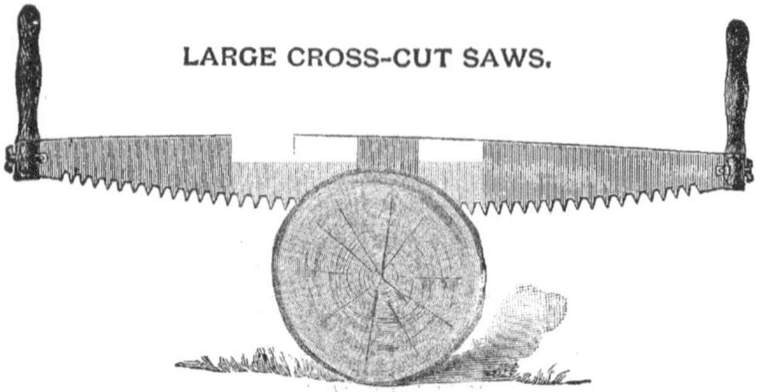

Fig. 19.

Fig. 19 represents a log of wood showing ends of grain which consist of more or less minute fibres or threads which constitute the tenacity of the wood. Our object with this saw is to sever the fibres or threads in the same manner as shown in cross-cut hand-saws, Figs. 3, 8, 10, 11 and 12, the only difference being that these large saws are constructed to cut equally well on either stroke, and that many of them have cleaner or drag teeth to carry out the dust, which we will consider later.

The same general rules for filing and setting cross-cut hand-saws apply to these saws, excepting that the angle of tooth is same on each side; the shape and space of teeth and different amounts of fleam, of course depends, as in other saws, largely upon the work to be performed. The greatest amounts of fleam are for the soft woods and less in proportion to hardness of same; this principle also applies to the amount of set.

Fig. 20.

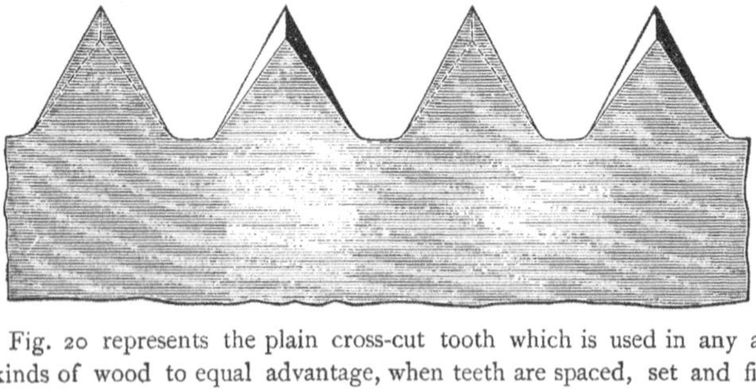

Fig. 20 represents the plain cross-cut tooth which is used in any and all kinds of wood to equal advantage, when teeth are spaced, set and filed in accordance with instructions given in the preceding pages in relation to proper space, pitch and bevel.

Fig. 21.

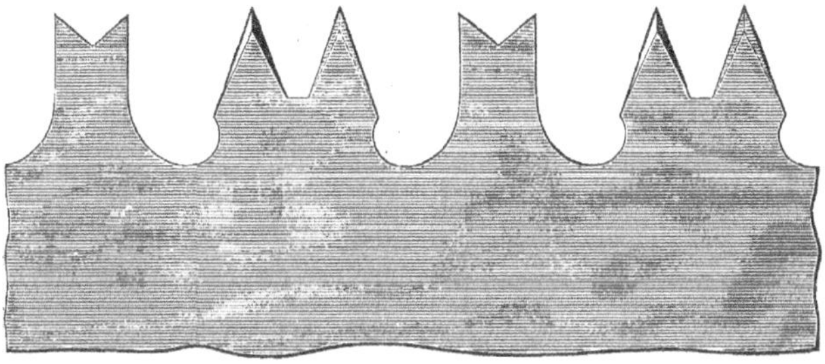

Fig. 21 represents a section of our improved California Tuttle tooth; this tooth is used principally on soft woods; the cleaner tooth must be filed square and about one thirty-second of an inch shorter than the cutting teeth.

Fig. 22.

Fig. 22 is a section of our Lumberman saw, showing the style of file that should be used to preserve the original shape of tooth. The style of setting and filing this saw is the same as the ordinary cross-cut hand-saw, each alternate tooth set and filed from reverse sides.

Fig. 23.

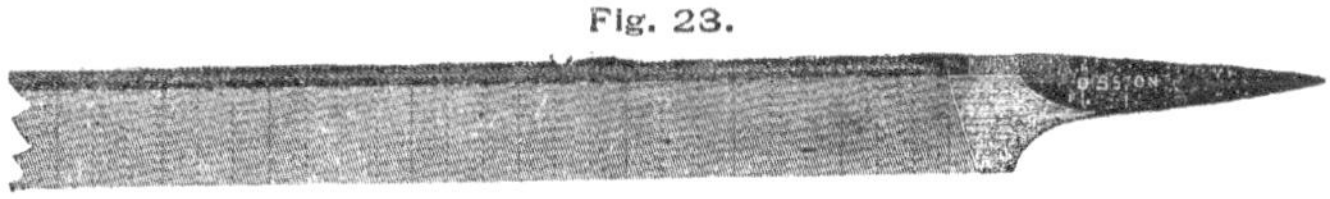

Fig. 23 represents file for keeping teeth of our Great American cross-cut saws in the same shape in which they leave our works.

Fig. 24.

Fig. 25.

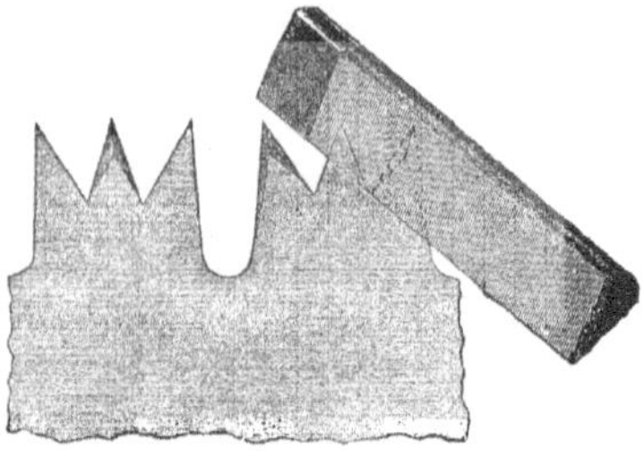

Fig. 24 shows the manner of filing the long edge of the end tooth.

Fig. 25 shows the manner of filing the short or inside edge of the end tooth.

Fig. 26.

Fig. 26 shows the section of the file in the gullet of the saw.

This file, though made expressly for our Great American cross-cut saw, will be found equally serviceable in filing the "Lumberman," "Climax," and other cross-cut saws.

The Great American tooth has been subjected to the most severe tesst, and is the best for general use ever offered to the public. These saws are ground extra thin back, which enables them to run with less set and more ease.

Fig. 27.

For filing the teeth as represented in Fig. 17, we make a special file represented above, by which the original shape of teeth is maintained. This style of tooth we put in our "Acme" saw, No. 120, which is a fast, smooth-cutting saw, and which runs entirely without set in dry seasoned lumber only. The teeth are made so hard, that an attempt to set them will inevitably result in breaking them.

BAND SAWS.

The only difference in the manner of setting and sharpening, and the shape of teeth of small band saws and ordinary hand rip-saws is that the teeth must have pitch enough to cause it to lead into the cut. In the ordinary rip-saw as shown in Fig. 2, the teeth meet the fibres of the wood at an angle which gives a shearing cut, while in the band saw, all the teeth strike the fibres alike—square across; consequently, if there is no pitch to the teeth, the saw is crowded back against the guide or guard, which not only subjects it to unnecessary strain, often breaking it, but crystallizes the back edge of blade from which cracks are sure to start. Many good saws are condemned and ruined from these two causes alone. Great care should be taken to guard against this, as well as against setting the teeth too deep, for if the set runs down to gullets of teeth, the blade will be distorted, give poor results, and eventually break at the roots.

Having explained the construction and methods of keeping saws in order, and enumerated a few of the many tools with which to do so, we will furnish our more complete price lists of our entire manufactures on application. Quality of goods considered, our prices are the lowest. Quotations promptly furnished and correspondence solicited.

Spotted something that isn't quite right?

Let us know and we'll fix it for future reprints.
And as a thank you for your diligence, we'll
send you a free Sothis book of your choice.

publisher@sothis.press

This page intentionally left blank

Made in United States
Cleveland, OH
06 November 2026

47780418R00059